*POLYSACCHARIDE GUMS FROM AGRICULTURAL PRODUCTS*

T0321136

# POLYSACCHARIDE GUMS FROM AGRICULTURAL PRODUCTS

## Processing, Structures & Functionality

# Steve W. Cui, Ph.D.

*Food Research Program*
*Agriculture and Agri-Food Canada*
*Guelph, Ontario*
*Canada*

CRC Press
Taylor & Francis Group
Boca Raton  London  New York

CRC Press is an imprint of the
Taylor & Francis Group, an **informa** business

CRC Press
Taylor & Francis Group
6000 Broken Sound Parkway NW, Suite 300
Boca Raton, FL 33487-2742

First issued in paperback 2019

ISBN-13: 978-1-56676-934-1 (hbk)
ISBN-13: 978-0-367-39799-9 (pbk)

Main entry under title:
    Polysaccharide Gums from Agricultural Products: Processing, Structures and Functionality

A Technomic Publishing Company book
Bibliography: p.
Includes index p. 261

Library of Congress Catalog Card No. 00-107624
ISBN No. 1-56676-934-5

**Visit the Taylor & Francis Web site at**
**http://www.taylorandfrancis.com**

**and the CRC Press Web site at**
**http://www.crcpress.com**

*To my wife Liqian Yao*

# Contents

# Preface

**T**HERE are only a few books available that address the basic chemistry, functional properties and applications of food hydrocolloids/gums. These books cover a wide spectrum of commercial gums and some potential gums, however, most of the research was done before the 1980s. Substantial progress has been made in the field of food hydrocolloids during the last two decades, and much research has been conducted on new gums. These new findings provide opportunities and guidelines for the applications of new gums in food systems, nevertheless, the information is scattered in the literature. There is an increased demand for books summarizing the information attained. This book meets this demand.

The basic chemistry, extracting process, molecular structure, and most importantly, functional properties and potential applications of new polysaccharide gums originating from agricultural products, such as yellow mustard seed, flaxseed, cereals (wheat, barley, oat and corn), legumes (soybean and fenugreek) and psyllium are discussed in this book. Most of these crops are grown in massive amounts, while some are used as or have potential to be used as alternative crops, such as psyllium and fenugreek. The utilization of these polysaccharides not only provides more choices of gums/stabilizers for the food industry, it also adds value to these crops.

# Acknowledgement

I would like to thank a number of people who have helped me as my career has developed, and who have, through their input, essentially enabled me to write this book. First, I would like to thank Professors Michael N. A. Eskin of the Department of Foods and Nutrition, University of Manitoba, Canada, and Costas Biliaderis of the Department of Food Science and Technology, Aristotle University, Greece, for introducing me to the field of food hydrocolloids. I would also like to thank my colleagues Dr. Peter Wood of the Food Research Program, Guelph, Ontario, and Dr. G. (Joe) Mazza of the Pacific Food Research Centre, Summerland, B.C., who have always been supportive and provided guidance and help and sometimes constructive discussions and suggestions. My sincere thanks are extended to Dr. Qi Wang for reviewing Chapter 3. Special thanks also go to my Director, Mr. Greg Poushinsky, for his support of my research program and the writing of this book. My thanks extend to Mrs. Cathy Wang who helped me redraw some of the figures, typed all of the tables and organized the references.

Finally, I would like to express my sincere and heartfelt thanks to my wife, Liqian, for her love, support and encouragement during the writing of this book.

# CHAPTER 1

# Yellow Mustard Gum

## 1. INTRODUCTION

**M**USTARDS have been consumed by humans as condiments for about 3000 years. The original use of mustard was to mask the taste of degraded perishables. Today, mustard is the largest volume spice in international trade, accounting for 160,000 tons per year [1]. As shown in Table 1.1, mustard seeds are composed of protein (23–30%), fixed oil (29–36%), carbohydrates (12–18%) and some minor constituents including minerals (4%), essential oil (glucosinolates, 0.8–2.3%) and phytin (2–3%) as well as some phenolic compounds and dithiolthiones. Processed mustard flour is usually enriched with fixed oil (30–42%) and protein (30–35%). In contrast, bran products contain much less oil and protein (7 and 13–16%, respectively), but are rich in dietary fiber (15%). Basic chemistry and functional properties of mustard components have been reviewed by Cui and Eskin [2].

This chapter will focus on the extraction process, chemical structure and functional properties of yellow mustard mucilage.

Traditional mustard products include mustard flour, ground mustard and prepared mustard. Mustard flour is a fine powder derived from the endosperm or interior portion of the seed. Three species of mustard seeds commonly ground into flour are yellow (*Sinapis alba*), oriental and brown mustards (*Brassica*

TABLE 1.1. Chemical Composition of Mustard Seed and Its Products.

| Mustard Product | Isothiocyanates | | Fixed Oil % | Protein[a] % | Crude Fiber % | H$_2$O % | Ash % |
|---|---|---|---|---|---|---|---|
| | Allyl- % | 4-Hydroxybenzyl- % | | | | | |
| Whole yellow seed | | 2.3 | 29 | 30 | 9 | 6 | 4 |
| Whole brown seed | 0.80 | | 32 | 26 | 7 | 6 | 4 |
| Whole oriental seed | 0.78 | | 36 | 23 | 6 | 6 | 4 |
| Yellow flour | | 0.0 | 30 | 35 | 3.5 | 6 | 4 |
| Brown flour | 0.95 | | 40 | 35 | 3.5 | 6 | 4 |
| Oriental flour | 0.90 | | 42 | 30 | 3.5 | 6 | 4 |
| Yellow bran | | | 7 | 16 | 15 | 10 | 4 |
| Brown bran | 0.20 | | 7 | 13 | 15 | 10 | 4 |
| Oriental bran | 0.35 | | 7 | 15 | 15 | 10 | 4 |

[a]Protein = $N \times 6.25$.
Reprinted from Reference [2] with permission from Technomic Publishing Co., Inc. © 1998.

*juncea*). Mustard flour is generally retailed directly or sold to industry as an ingredient for such products including salad dressings, mayonnaise, barbecue sauce, pickles and processed meats. Ground mustard, a powder made by grinding whole yellow mustard seeds, is used primarily in the meat industry as an emulsifier, water binder and inexpensive bulking agent. It is also used in seasonings for frankfurters, bologna, salami, lunch loaf, salad dressings and pickled products. The water binding and emulsifying properties of ground mustard are largely attributed to the mucilaginous material present in yellow mustard bran. All mustard seeds contain mucilage, but only yellow mustard mucilage is significant because of its high yield and unique functional properties [3].

Yellow mustard bran used to be a by-product in the preparation of mustard flour. Recently, the bran has been in higher demand than the flour because of its excellent emulsifying and stabilizing properties. However, only a few researchers have paid attention to the chemical structure and functional properties of yellow mustard mucilage. The earliest report, by Bailey and Norris in 1932, studied the basic chemical structure of yellow mustard mucilage [4]. In the 1960s and 1970s, studies were carried out by Rees and coworkers concerning the relationships between polysaccharide structures and germination properties of yellow mustard seeds (which were referred to as white mustard seeds) [5]. Weber and coworkers (1974) reported preliminary functional properties of yellow mustard mucilage and its synergistic interactions with locust bean, guar gum and carboxymethylcellulose [3]. In the 1980s, two groups from Canada showed interest in yellow mustard mucilage because Canada is the largest producer of yellow mustard [6,7]. During the last decade, Cui and Eskin carried out a systematic study covering basic chemical structures, rheological properties and applications in food systems of yellow mustard mucilage [2]. This chapter reviews the most recent advances in the area. To be consistent with other commercial polysaccharide gums, yellow mustard mucilage will be referred to as yellow mustard gum (YMG) in this chapter.

## 2. OCCURRENCE AND EXTRACTION PROCESS

### 2.1. OCCURRENCE

When yellow mustard seeds are exposed to water, the surfaces of the seeds become sticky. If the seeds are dried immediately after wetting, some of them will glue together. This phenomenon can be easily explained by the presence of a material called "mucilage" in the seed coat of yellow mustard seed. The mucilage is deposited in the epidermal layer of the seed coat, and it is readily exuded in a high moisture environment [7,8]. The release of mucilage from the seeds when immersed in water can be observed with the naked eye. If the dampened seeds are allowed to dry, the exuded mucilage forms a whitish layer on the surface. Using optical and scanning electron microscopy, high quality

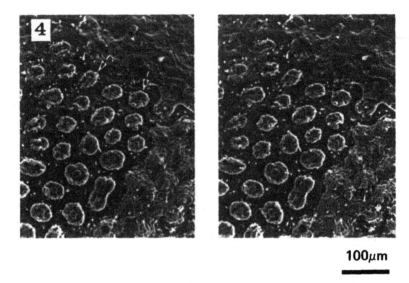

**100μm**

**Figure 1.1** Mucilage emerged from the hull of a seed moistened for 1 hr and subsequently dried. The mucilage is in the form of individual minute droplets (arrows) or in the form of droplets coalesced into a stratified cover (C). In this stereo pair of micrographs (12° angular separation), two dots have been provided to facilitate focusing the eyes. Reprinted from Reference [7].

micrographs of mucilage exuded from moistened seeds and subsequently dried mucilage in the form of small droplets on the seed surface of yellow mustard have been obtained (Figure 1.1).

## 2.2. EXTRACTION FROM WHOLE SEEDS

Because the mucilage is deposited on the epidermal layer of the testa, it is relatively easy to extract with water. Bailey and Norris used cold water to extract the mucilage from yellow mustard seed [4]. The seed to water ratio is 1:10 with efficient stirring or shaking for 24 hr. The viscous extract is filtered through muslin, while the seed is returned to the vessel for two additional extractions. The three extracts are combined and then filtered through a bed of glass wool and muslin to remove foreign particles. The filtrate is then poured into alcohol to precipitate the mucilage. The fibrous, gelatinous material from alcohol precipitation is redissolved, precipitated and filtered several times, and the final product is dried in successive increasing concentrations of alcohol and finally under vacuum over phosphorus pentoxide [4]. The prepared mucilage is a snow-white fibrous product with a yield of ~2%. In an extraction procedure reported by Woods and Downey, a small amount of chloroform was added to water (2.5 mL/L) to prevent undesirable fermentation during the long extraction period [6]. The extraction is carried out by shaking the whole seed (5 g) slowly in 90 mL of a chloroform-water mixture overnight (16 hr). The

TABLE 1.2. **Percent Mucilage in the Seed of Four Yellow Mustard Cultivars Grown at Four Locations Over a Two Year Period.**

| Location | Average (%) | Cultivar | Average (%) | Year | Average (%) |
|----------|-------------|----------|-------------|------|-------------|
| Melfort | 2.15 | BHL 3-926 | 1.94 | 1977 | 1.79 |
| Sidney | 1.76 | Sabre | 1.55 | 1978 | 1.41 |
| Winnipeg | 1.31 | Gidilba | 1.49 | | |
| Saskatoon | 1.17 | Yellow-2 | 1.41 | | |

LSD 5%, locations and cultivars 0.37%, year 0.26%.
Reprinted from Reference [6] with permission from the *Canadian Journal of Plant Science.* © 1980 Agricultural Institute of Canada.

seeds are removed by straining them through cheesecloth, and the mucilage is precipitated in acidified acetone (2.5 mL conc. HCl/L) and dried under vacuum at room temperature. The yield is significantly affected by maturity, cultivars, growing years and locations, as shown in Table 1.2 [6]. The relative low yields reported in Woods and Downey's work (0.3–2.0%) might be due to incomplete extraction. Single extraction was used compared to three sequential extractions by Bailey and Norris [4]. A yield of 5% was reported by Cui and coworkers [9], which is significantly higher than that reported by Bailey and Norris [4] and by Woods and Downey [6]. In this procedure, 100 g of yellow mustard seeds are poured into 600 mL of boiling water. The temperature of the sample is kept at 75°C in a water bath for 30 min. The sample is allowed to cool to room temperature, and the extraction is continued overnight (16 hr) at room temperature with constant stirring using a magnetic stirring bar. Chloroform (2.5 mL/L) is added to prevent possible fermentation [6]. The extracted solution is filtered through three-layer cheesecloth. After washing with 50 mL of water, the seed retained is subjected to two additional extractions. The three extracts are combined and precipitated in three volumes of 95% ethanol. A fibrous, gelatinous product formed from the alcohol precipitation is recovered by centrifugation, washed three times with 70% ethanol, then redissolved in water and freeze dried. The final mucilage product obtained is snow-white and cotton-like in appearance, similar to the mucilage extracted by Bailey and Norris [4]. The three successive extractions yielded 5% mucilage, accounting for 93% of the total extractable under the described conditions (Table 1.3).

TABLE 1.3. **Yield of Mucilage from Yellow Mustard Seeds upon Sequential Aqueous Extraction (100 g seeds/600 g $H_2O$).**

| Extraction | 1 | 2 | 3 | 4 | 5 | Total |
|------------|-----|-----|-----|-----|-----|-------|
| Weight (g) | 3.45 | 0.93 | 0.54 | 0.25 | 0.12 | 5.29 |
| Percent (%) | 65.2 | 17.6 | 10.2 | 4.7 | 2.3 | 100 |

Reprinted from Reference [9] with permission from Elsevier Science.

## 2.3. EXTRACTION OF YMG FROM BRAN

Using a whole seed as a source of YMG is not economically feasible because there is no practical use for the seeds after the extraction. In North America, substantial amounts of mustard meats are used as food ingredients, while the bran is considered a by-product. Because mucilage is deposited in the epidermal layer of the seed coat, extraction of mucilage from the bran could be commercially viable for reducing the processing cost and for allowing utilization of the by-product. The yield of YMG from bran varied from 15–25% depending on the extraction conditions [3,5,10]. The extraction process generally includes (1) defatting the bran with a mixture of hexane, ethanol and water; (2) extracting the bran with water (1:20 ratio) and separating by centrifugation; and (3) precipitating in ethanol, followed by drying (freeze drying or under vacuum). Siddiqui and coworkers extracted YMG from yellow mustard bran with boiling water (1:16, w/v) for 35 min [7]. The extract is separated by centrifugation then precipitated by adding isopropanol to a final concentration of 70% (v/v). The precipitate is filtered by an organdy cloth, washed with 70% isopropanol, air-dried and pulverized [7].

Cui and coworkers recently optimized an extraction process using Response Surface Methodology [11]. In a center composite design, temperature, pH, water/solid ratio and extraction time were the four independent factors examined at five levels; two dependent responses, yield and the apparent viscosity of extracted gum in an aqueous solution were determined to evaluate the optimization process. Of the four independent factors examined, temperature and pH were found to have a much greater influence on the yield and rheological properties of the extracted gum compared to water/solid ratio and extraction time. Optimum extraction conditions are temperatures between 50–55°C, pH 8.5–9.5, water/solid ratio of 50–55 and extraction time of 2–2.5 hr, as shown in Figure 1.2. The yield of gum obtained under optimum conditions is 30% of bran weight, and the gum exhibits shear thinning flow behavior comparable to the mucilage extracted from the whole seed [9].

## 3. CHEMICAL COMPOSITION, STRUCTURES AND MOLECULAR WEIGHT DISTRIBUTION OF YMG

### 3.1. CHEMICAL AND MONOSACCHARIDE COMPOSITIONS

Yellow mustard mucilage was first reported to be composed of cellulose (~50%) and acidic polysaccharides containing arabinose, galactose, rhamnose, galacturonic acid and glycuronic acid with methoxyl groups [4]. A separate study suggested that galactose, arabinose and galacturonic acid were the major constituent monosaccharides, whereas xylose, glucose, mannose and rhamnose

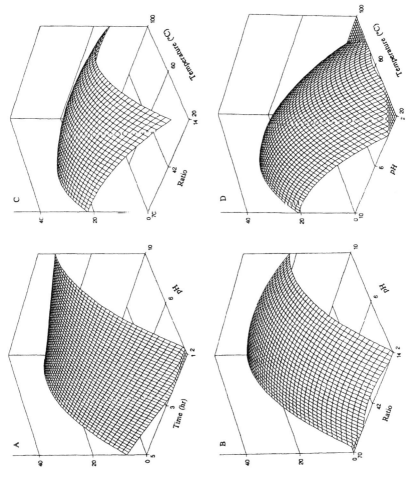

**Figure 1.2** Effect of extraction conditions on the apparent viscosity of yellow mustard gum. Reprinted from Reference [11].

**TABLE 1.4. Chemical Compositions of Yellow Mustard Seed Mucilage and Its Fractions.**

| Component (on Dry Base) | | CMa | CMb | WS | WI |
|---|---|---|---|---|---|
| Water | (%) | 6.9 | 8.7 | 10.2 | 8.1 |
| Ash | (%) | 15.0 | 4.8 | 4.3 | 1.7 |
| Protein | (%)c | 4.4 | 4.1 | 2.2 | 4.2 |
| Fat | (%) | 0.2 | ...d | ... | ... |
| Carbohydrate | (%)e | 80.4 | 91.1 | 93.5 | 94.1 |
| Potassium | (%) | 2.1 | 0.04 | 0.01 | 0.02 |
| Calcium | (%) | 2.2 | 1.2 | 1.6 | 0.48 |
| Magnesium | (%) | 1.3 | 0.6 | 0.4 | 0.16 |
| Phosphorus | (%) | 2.1 | 0.6 | 0.4 | 0.09 |
| Sulphur | (%) | 1.4 | 0.29 | 0.16 | 0.02 |
| Iron | (ppm) | 212.5 | 260.0 | 223.0 | 358.0 |
| Zinc | (ppm) | 53.5 | 70.0 | 123.0 | 208.0 |
| Manganese | (ppm) | 73.0 | 70.0 | 72.0 | 27.0 |
| Copper | (ppm) | 13.5 | 12.0 | 34.0 | 18.0 |

aCrude mucilage before dialysis.
bCrude mucilage after dialysis.
c$N \times 6.25$.
dNot determined.
eBy difference.
CM = mucilage; WS = water-soluble fraction; WI = water-insoluble fraction.
Reprinted from Reference [9] with permission from Elsevier Science.

were the minor constituents [5]. A quantitative analysis revealed that YMG contained 39.3% glucose, 25.4% arabinose, 17.9% galactose, 7.5% xylose, 5.4% mannose, 4% rhamnose and 1% fructose [12]. Other studies on the composition of YMG were generally in agreement with previous results, but there were discrepancies [7,10,12]. Because previous studies are inconsistent, a systematic study on the extraction, fractionation and physicochemical characterization of YMG was carried out by Cui and Eskin [13–17]. The crude gum (CM) extracted by Cui and coworkers contained 80.4% carbohydrates, 4.4% protein and 15% ash. Dialysis reduced the ash content to 4.8% with a corresponding increase in carbohydrates from 80.4 to 91.1% (Table 1.4). Of the monosaccharides, glucose (23.5%) was the predominant neutral sugar, followed by galactose (13.8%), mannose (6.1%), rhamnose (3.2%), arabinose (3.0%) and xylose (1.8%) together with 14.7% uronic acids.

## 3.2. HETEROGENEITY AND FRACTIONATION OF YMG

YMG is a mixture of complex polysaccharides containing six neutral sugars and two uronic acids as described in section 3.1. The heterogeneity of YMG

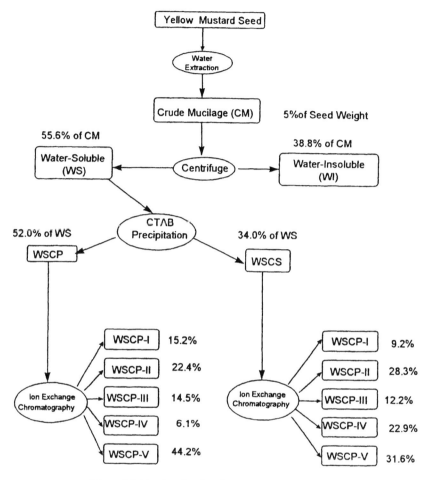

**Figure 1.3** Fractionation flowchart of yellow mustard gum.

is not only reflected in monosaccharide composition, but also in differences in molecular weight distribution. As described in section 2.2, YMG is extracted as crude mucilage from the seed or bran. The crude mucilage (CM) can be fractionated into a number of fractions. A flowchart presented in Figure 1.3 shows the fractionation procedures and all fractions obtained. First, CM is separated into a water-soluble (WS) fraction and a water-insoluble (WI) fraction by centrifugation. The WS fraction was identified as the major fraction and exhibited shear thinning flow behaviors observed for YMG [9]; therefore, the focus of the research was on the WS fraction. There are at least two types of polysaccharides present in the WS fraction: a neutral polysaccharide mainly composed of glucose and an acidic polysaccharide comprised of two uronic acids, galactose and rhamnose. A quaternary ammonium cation (hexadecyltrimethylammonium

bromide, also noted as CTAB) is used to further fractionate the WS fraction because CTAB can form a complex (precipitate) with the acidic polysaccharides.
As a result, two fractions are obtained from WS: a CTAB-precipitated (WSCP)
and a CTAB-soluble (WSCS) fraction [14]. When CTAB was used in an excess amount, all polysaccharides, including the neutral fraction, were precipitated. An optimum CTAB concentration was selected to give the two fractions;
as a result, two or more polysaccharides were identified from both fractions
(Table 1.5). Nevertheless, the content of uronic acids in WSCP (23%) was
significantly higher than that in WSCS (12.5%). Two different acidic polysaccharides must be present in WS: one has more acidic groups and can easily form
precipitate with CTAB (WSCP), while the other contained less acidic groups
and was soluble under the CTAB concentration used. The co-precipitation of
the neutral polysaccharides together with the weak acidic polymers indicates
that there is an association/interaction between the neutral and weak acidic
polysaccharides. WSCP and WSCS were further fractionated by ion-exchange
chromatography, which yielded five subfractions from each fraction, as shown
in Figure 1.3. A sample in 1.0% (w/w) solution was introduced to the column
and eluted with appropriate ion strength of NaAc buffer (pH 5.0), followed by
the elution of the same buffer until no carbohydrate is detected by the phenol-
sulphuric acid method [15]. The ionic strength of the elution buffer (pH 5.0) is
increased stepwise to obtain five fractions. The WSCP series includes WSCP-I
(neutral fraction), WSCP-II (0.2 M NaCl buffer elution), WSCP-III (0.5 M
NaCl buffer elution), WSCP-IV (1.0 M NaCl buffer elution) and WSCP-V
(6 M urea elution). In the WSCS series, a much weaker ionic strength buffer
system was used: a weak ionic strength like 0.2 M NaCl eluted almost all of the
acidic polysaccharides in WSCS, however, 1.0 M NaCl is required to remove
all acidic polysaccharides from the column.

The yield, monosaccharide composition and content of uronic acids of WS
and WI are presented in Table 1.5. The water-soluble fraction contained higher
amounts of uronic acid and corresponding higher levels of galactose and rhamnose compared to the insoluble fraction. In contrast, the glucose content of WI
(35%) is higher than that of the CM (23.5%) (Table 1.5).

The structural heterogeneity of the fractions is revealed by methylation
analysis, as shown in Tables 1.6 and 1.7. In the WSCP series, WSCP-I contained
44% (1→4)-β-D-glucosyl residues and 18.7% (1→4)-β-xylosyl residues with
minor amounts of arabinosyl, galactosyl and mannosyl residues, but no uronic
acids were detected (Table 1.6). WSCP-V only contained β-D-glucosyl residues,
mostly in the (1→4) linkage. The variations in content of the two uronic acids
identified in the acidic polysaccharide fractions (glucuronic acid and galacturonic acid) indicate heterogeneity in the structure of acidic polysaccharides. Of
the three acidic polysaccharide fractions, WSCP-III appeared to be pure with
only a trace amount of cellulose-like material. A rhamnogalacturonan structure
was suggested that is highly branched with galactose and glucuronic acid, as

TABLE 1.5. Yield, Monosaccharide Composition and Uronic Acid Content of Yellow Mustard Mucilage and Its Fractions.

| Mucilage and Fraction | Sample Yield (%) | Uronic Acid (%) | Glucose (%) | Galactose (%) | Mannose (%) | Rhamnose (%) | Arabinose (%) | Xylose (%) |
|---|---|---|---|---|---|---|---|---|
| CM | ... | 14.64 | 23.54 | 13.83 | 6.07 | 3.15 | 3.02 | 1.80 |
| WS | 55.6 | 18.68 | 22.26 | 15.21 | 6.31 | 3.93 | 3.22 | 1.77 |
| WI | 38.8 | 10.30 | 34.95 | 11.70 | 6.35 | 1.65 | 2.84 | 2.00 |

CM = crude mucilage; WS = water-soluble fraction; WI = water-insoluble fraction.
Reprinted from Reference [9].

11

YELLOW MUSTARD GUM

**TABLE 1.6.** Molar Ratios of Partially Permethylated Acetyl Alditols
of the Water-Soluble CTAB-Precipitated (WSCP) Fraction
of Yellow Mustard Gum.

| | Molar Ratio (%)[a] | | | | |
|---|---|---|---|---|---|
| | WSCP-I | WSCP-II[b] | WSCP-III | WSCP-IV | WSCP-V |
| 2,3,5-Me$_3$-Ara | 2.8 | 0.8 | 0.4 | 3.6 | 0 |
| 2,3-Me$_2$-Ara | 4.3 | 2.7 | 2.5 | 4 | 0 |
| Total methyl ethers of Ara | 7.1 | 3.5 | 2.9 | 7.6 | 0 |
| 2,3-Me$_2$-Xyl | 18.7 | 4.0 | 1.3 | 1.8 | 0 |
| Xyl (Acet)$_5$ | 3.9 | 0.6 | tr | 15.3 | 0 |
| Total methyl ethers of Xyl | 22.6 | 4.6 | 1.3 | 17.1 | 0 |
| 2,3,4,6-Me$_4$-Glc | 1.6 | 1.4 | 0 | 4.2 | 1.8 |
| 2,3,6-Me$_3$-Glc | 44.5 | 13.9 | 1.0 | 4.3 | 81.3 |
| 2,3-Me$_2$-Glc | tr | 5.5 | 0 | 1.7 | 0 |
| Total methyl ethers of Glc | 46.1 | 20.8 | 1.0 | 10.2 | 83.1 |
| 2,3,4-Me$_3$-Glc | n.d. | 15.9 | 13.3 | n.d. | n.d. |
| 2,3,4,6-Me$_4$-Gal | n.d. | 5.5 | 2.3 | 4.0 | 0 |
| 3,4,6-Me$_3$-Gal | 3.0 | 0 | 2.4 | 4.6 | 0 |
| 2,3,4-Me$_3$-Gal | 5.9 | 23.8 | 22.9 | 39.4 | 0 |
| 2,4-Me$_2$-Gal | tr | 0.8 | 0.6 | tr | 0 |
| Total methyl ethers of Gal | 8.9 | 30.1 | 28.2 | 48.0 | 0 |
| 2,3-Me$_2$-Gal (6D2) | n.d. | 5.4 | 13.6 | 5.6 | n.d. |
| 2,3,6-Me$_2$-Man | 4.5 | 5.3 | 0.2 | 0 | 0 |
| 3,4-Me$_2$-Rham | 1.2 | 2.1 | 11.0 | 0 | 0 |
| 3-Me-Rham | tr | 5.6 | 17.7 | 4.9 | 0 |
| Total methyl ethers of Rham | 1.2 | 7.7 | 28.7 | 4.9 | 0 |

[a]Relative molar ratio calculated from the ratio of peak heights.
[b]WSCP-II, III and IV were carboxyl reduced.
n.d.: not determined.
tr: trace amount.
Reprinted from Reference [14] with permission from Elsevier Science.

will be discussed in detail in section 3.4. WSCP-II and WSCP-IV still contained
substantial amounts of cellulose-like material (Table 1.6). In the WSCS series,
WSCS-I and WSCS-V were similar in composition and linkage pattern, primarily comprised of $(1\rightarrow4)$-$\beta$-D-glucosyl residues (Table 1.7). However, the
monosaccharide composition and linkage patterns of WSCS-II differ from those
of corresponding fractions of WSCP by containing significantly higher amounts
of mannose, arabinose and xylose. WSCS-I and WSCP-III were identified exhibiting the shear thinning flow behavior observed for YMG and were relatively

TABLE 1.7. Molar Ratios of Partially Methylated Alditol Acetates
of the Water-Soluble CTAB-Precipitated (WSCS) Fractions
of Yellow Mustard Mucilage.

| | Molar Ratio[a] | | | | |
|---|---|---|---|---|---|
| | WSCS-I | WSCS-II[b] | WSCS-III | WSCS-IV | WSCS-V |
| 2,3,5-Me$_3$-Ara | 1.8 | 1.0 | 0.6 | tr | tr |
| 2,3-Me$_2$-Ara | 1.6 | 5.8 | 6.2 | tr | 1.2 |
| Total methyl ethers of Ara | 3.4 | 6.8 | 6.8 | . . . | 1.2 |
| 2,3-Me$_2$-Xyl | tr | 2.0 | 5.2 | tr | 1.0 |
| Xyl (acet) | 2.3 | 0.3 | 0.3 | tr | 2.0 |
| Total methyl ethers of Xyl | 2.3 | 2.3 | 5.5 | . . . | 3.0 |
| 2,3,4,6-Me$_4$-Glc | 3.2 | 2.4 | 4.2 | 0 | 1.8 |
| 2,3,6-Me$_3$-Glc | 66.0 | 17.5 | 4.9 | tr | 71.6 |
| 2,3-Me$_2$-Glc | 4.1 | 8.4 | 4.9 | 0 | 5.2 |
| Total methyl ethers of Glc | 73.3 | 27.9 | 14.0 | . , , | 78.6 |
| 2,3,4-Me$_3$ Glc (GD$_2$) | n.d. | 9.0 | 9.7 | 6.0 | n.d. |
| 2,3,4,6-Me$_4$-Gal | 1.6 | 2.0 | 4.0 | tr | 0.4 |
| 3,4,6-Me$_3$-Gal | 0.8 | 8.8 | 2.1 | tr | 0.5 |
| 2,3,4-Me$_3$-Gal | 1.4 | 12.4 | 12.1 | 21.8 | 1.5 |
| 2,4-Me$_2$-Gal | 6.9 | 0.4 | 2.8 | tr | 7.6 |
| Total methyl ethers of Gal | 10.7 | 23.6 | 21 | 21.8 | 10.0 |
| 2,3-Me$_2$-Gal (6D$_2$) | n.d. | tr | 3.3 | 2.4 | n.d. |
| 2,3,6-Me$_3$-Man | 1.2 | 8.7 | 1.3 | 0 | 1 |
| 3,4-Me$_2$-Rham | 0.6 | 1.2 | 3.4 | tr | 0 |
| 3-Me-Rham | tr | 0.4 | 4.7 | tr | 1.1 |
| Total methyl ethers of Rham | 0.6 | 1.6 | 8.1 | . . . | 1.1 |

[a]Relative molar ratio calculated from the ratio of peak heights.
[b]WSCS-II, III and IV were carboxyl reduced.
n.d.: not determined.
tr: trace amount.
Reprinted from Reference [14] with permission from Elsevier Science.

pure. Detailed structural analyses were carried out for these two fractions, as is detailed in the next two sections.

## 3.3. STRUCTURE OF WATER-SOLUBLE (1→4)-LINKED β-D-GLUCAN (WSCS-I)

Methylation analysis revealed that WSCS-I is comprised of 66% (1→4)-β-D-glucosyl backbone chain with 4% 6-substituted (1→4)-β-glucosyl linkage.

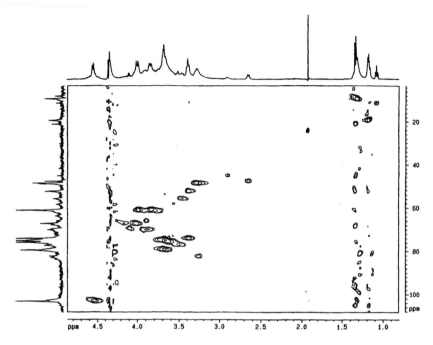

**Figure 1.4** $^1H/^{13}C$ heteronuclear correlation NMR spectrum of a water-soluble 1,4-linked β-D-glucan from yellow mustard mucilage (4% polymer in $D_2O$, 65°C). Reprinted from Reference [17] with permission from Elsevier Science.

A combination of one- and two-dimensional and $^{13}C/^1H$ NMR spectroscopy confirmed that some of the hydroxyl groups at C2, 3 and 6 are substituted by ether groups (ethyl and propyl) [17]. The ethyl group was found randomly distributed in the C2, 3 and 6 positions, while the propyl ether was predominant at the 6 position. These ether groups along the cellulose-like backbone chain of WSCS-I may act as "kinks" that alter the conformational regularity of the (1→4)-linked β-D-glucosyl backbone chain and favor the solubilization of the polymer in aqueous mediums. Alternatively, for steric reasons, these groups will hinder interchain associations among the cellulose chains, thus enhancing solubility of the polysaccharide.

In the $^{13}C$ NMR spectrum of WSCS-I, six major signals were observed for the corresponding six carbons of the repeating (1→4)-β-D-glucosyl residues: 103.08 ppm, C1; 73.84 ppm, C2; 74.95 ppm, C3; 79.36 ppm, C4; 75.67 ppm, C5; and 60.96 ppm, C6 (projection of Figure 1.4). In addition to the signals derived from carbohydrates, there were three groups of minor resonances detected in the regions of 9–21 ppm, 45–60 ppm and 65–70 ppm, respectively. This observation suggests that WSCS-I may have some branches. The observed non-sugar resonances are probably caused by ether, ester or cyclic acetal derivatives of cellulose that are not common in naturally occurring celluloses. The absence

TABLE 1.8. **Complete Assignment of the Major Resonances of $^1$H and $^{13}$C Spectra of WSCS-I.**

|  | 1 (C, H) | 2 (C, H) | 3 (C, H) | 4 (C, H) | 5 (C, H) | 6 (C, H) |
|---|---|---|---|---|---|---|
| C (ppm) | 103.08 | 73.84 | 74.95 | 79.36 | 75.67 | 60.96 |
| H (ppm) | 4.54 | 3.38 | 3.65 | 3.66 | 3.52 | 3.82, 3.98 |
|  | $J_{1,2}$ 7.5 Hz |  |  |  |  |  |

$J_{1,2}$. Coupling constant of protons at 1 and 2 positions.
Reprinted from Reference [17] with permission from Elsevier Science.

of signals in the region of 150–190 ppm suggested that those non-sugar signals were not caused by esters because esters should have strong carbonyl signals in this region. Resonance in the $^1$H NMR spectrum of WSCS-I appeared to be difficult to assign due to poor resolution and heavy overlap in the resonance region for sugars (3–4.5 ppm); three strong signals at 1–1.5 ppm, however, can be attributed to -CH$_3$ groups of the ethers.

A $^1$H-$^{13}$C correlation experiment was conducted to resolve the overlap problem (Figure 1.4). Spectrum resolution was significantly improved because the signals were spread out over the two-dimensional domain. Chemical shift at 4.54 ppm in the $^1$H spectrum correlated to a signal at 103.08 ppm of the $^{13}$C spectrum and, therefore, is attributed to the anomeric proton (C-1). In a similar manner, all signals in the $^1$H spectrum are assigned to their corresponding counterparts in the $^{13}$C spectrum. A complete assignment of the resonances of the $^1$H and $^{13}$C NMR spectra are shown in Table 1.8 (sugar ring $^1$H and $^{13}$C resonances) and Table 1.9 (non-sugar $^1$H and $^{13}$C resonances), respectively. The $^1$H-$^1$H coupling constant of the anomeric proton was 7.5 Hz (Table 1.8)

TABLE 1.9. **$^1$H/$^{13}$C Correlations and Assignment of Non-Sugar Resonances.**

| $^1$H (ppm) | $^{13}$C (ppm) | Assignment[a] | Position Assigned on Glucose Ring |
|---|---|---|---|
| 1.07 | 11.0 | a: **CH$_3$**-CH$_2$-O | 3 |
| 1.17 | 19.22 | b: **CH$_3$**-CH$_2$-O | 2 |
| 1.32 | 21.0 | c: **CH$_3$**-CH$_2$-O | 6 |
| 1.32 | 9.04 | d: **CH$_3$**-CH$_2$-CH$_2$-O | 6 |
| 2.65 | 47.65 | a: CH$_3$-**CH$_2$**-O | 3 |
| 3.27 | 48.5 | d: CH$_3$-**CH$_2$**-CH$_2$-O | 6 |
| 3.98 | 66.96 | b: CH$_3$-**CH$_2$**-O | 2 |
| 4.05 | 69.5 | c: CH$_3$-**CH$_2$**-O | 6 |
| 3.88 | 69.79 | d: CH$_3$-CH$_2$-**CH$_2$**-O | 6 |

[a]a, b, c and d correspond to different ether groups while the groups in bold indidate the proton and carbon which are responsible for the assigned resonances.
Reprinted from Reference [17] with permission from Elsevier Science.

which confirmed that the anomeric carbon was in β configuration. As shown in Table 1.9, non-sugar signals were classified into three groups according to their chemical shift (Group 1: $^1$H 1.07–1.32 ppm, $^{13}$C 9–21 ppm; Group 2: $^1$H 2.65–3.4 ppm, $^{13}$C 44–55 ppm; and Group 3: $^1$H 3.2–4.1 ppm, $^{13}$C 63–70 ppm). The signals in the first group were attributed to -CH$_3$ of the ether moiety; signals in the second group could be caused by -CH$_2$- in the propyl and/or -CH$_2$-O of ethyl at position 3 of some of the glucosyl residues. Signals in the third group could arise from -CH$_2$-O of the ethyl and propyl moieties.

$^1$H-$^1$H homonuclear correlation spectroscopy (COSY) was carried out to establish the connectivity of the signals. The intra-residue connectivity of the sugar ring starts from H1 → H2 → H3 → H4 → H5 → H6, as shown in Figure 1.5.

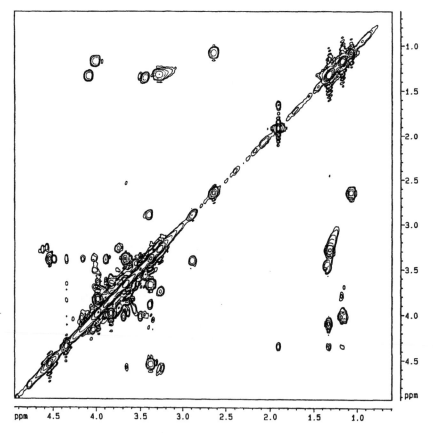

**Figure 1.5** Homonuclear shift correlated spectrum (COSY) of a water-soluble 1,4-linked β-D-glucan from yellow mustard mucilage (4% polymer in D$_2$O, 65°C). Reprinted from Reference [17] with permission from Elsevier Science.

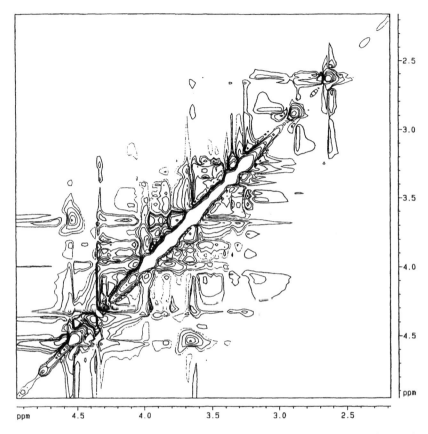

**Figure 1.6** NOESY of a water-soluble 1,4-linked β-D-glucan from yellow mustard mucilage (4% polymer in $D_2O$, 65°C). Reprinted from Reference [17] with permission from Elsevier Science.

The connectivity of H3 and H4 cannot be distinguished because the two signals are too close to each other (3.65 and 3.66 ppm, respectively). Four ether groups were identified and their connectivities were established for the non-sugar signals as shown in Table 1.9. Ether groups a, b and c have the same chemical formula ($CH_3CH_2$-O-) but are different in chemical shifts in $^1H$ and $^{13}C$ NMR spectra. These changes in chemical shift may suggest their attachment to different sites of the glucosyl residues (i.e., C2, C3, C6).

The NOESY spectrum of WSCS-I, shown in Figure 1.6, provided information on the intra- and inter-residue connectivities based on dipole correlation (through space). Strong correlation was observed between $^1H1$ and $^1H4$, an inter-residue correlation that confirmed (1→4)-linkage of the glucosyl residues. From the NOESY spectrum, correlations were also observed between ethyl **a** and $^1H3$; ethyl **b** and $^1H2$; ethyl **c** and $^1H6$; and propyl ether **d** and $^1H6$. This

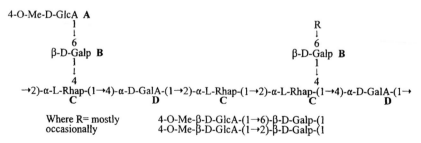

**Figure 1.7** Model structure of $(1 \rightarrow 4)$-β-D-glucan with ethyl and propyl substitutes: $R_1$ = propyl group; $R_2$ = ethyl group.

observation suggested that the ethyls were linked at 2, 3 and 6 positions, while only the propyl could be found at 6 position on the glucosyl residues. A model structure of WSCS-I is shown in Figure 1.7.

## 3.4. STRUCTURE OF A RHAMNOGALACTURONAN (WSCP-III)

An average repeating structure unit of WSCP-III, elucidated from methylation analysis, one- and two-dimensional NMR spectroscopy and characterization of oligosaccharides released from partial hydrolysis, is shown in Figure 1.8.

Methylation analysis revealed that the purified WSCP-III comprised of a terminal nonreducing D-glucuronic acid, 4-linked-D-galacturonic acid, 6-linked-D-galactose, and 2,4-linked and 2-linked L-rhamnose in the ratio of 0.9:0.8:1.5:1:0.5 [16].

Four signals were observed in the anomeric region of the $^{13}$C NMR spectrum of WSCP-III, which suggested the presence of at least four anomeric carbons in the polymer (projection of Figure 1.9). This observation is in agreement with the result from methylation analysis in which two neutral sugars and two uronic acids were identified. Signals at 176.22 and 175.39 ppm originated from the carboxyl groups of the terminal nonreducing D-glucuronic acid and 4-linked D-galacturonic acid, respectively. The signal at 60.35 ppm was attributed to

```
4-O-Me-D-GlcA  A                                              R
               1                                              1
               ↓                                              ↓
               6                                              6
           β-D-Galp  B                                    β-D-Galp  B
               1                                              1
               ↓                                              ↓
               4                                              4
→2)-α-L-Rhap-(1→4)-α-D-GalA-(1→2)-α-L-Rhap-(1→2)-α-L-Rhap-(1→4)-α-D-GalA-(1→
          C               D               C               C               D
```

Where R= mostly      4-O-Me-β-D-GlcA-(1→6)-β-D-Galp-(1
       occasionally   4-O-Me-β-D-GlcA-(1→2)-β-D-Galp-(1

**Figure 1.8** A repeating unit of a rhamnogalacturonan from yellow mustard gum.

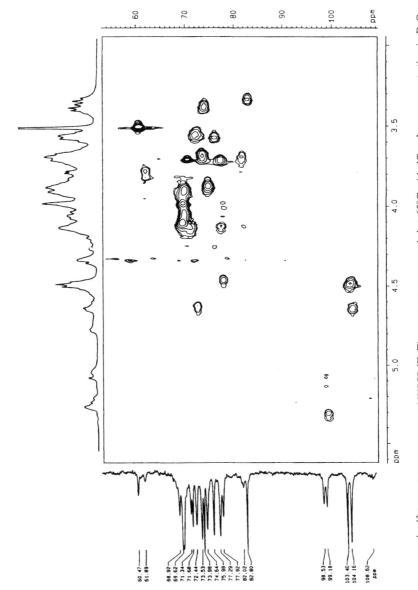

**Figure 1.9** $^1$H/$^{13}$C correlation spectrum of WSCP-III. The spectrum was recorded at 65°C with 4% polymer concentration in D$_2$O. Reprinted from Reference [16] with permission from Elsevier Science.

an $O$-methyl group attached at the 4 position of the D-glucuronic acid. Two signals at high field, 17.85 and 17.50 ppm, were in an approximate ratio of 2:1 in resonance intensity, which is attributed to the -$CH_3$ (C-6) of the 2,4-linked and 2-linked rhamnose, respectively. This ratio is in agreement with the result of methylation analysis of the two residues. A complete assignment of the [1]H-[13]C spectra of WSCP-III is achieved by [1]H-[13]C heteronuclear correlated spectroscopy and COSY and by comparison of the observed data with literature values, as presented in Table 1.10.

A full assignment of the [1]H and [13]C signals of the 4-$O$-Me-β-D-glucuronic acid (A in Figure 1.8) is summarized in Table 1.10. H-1 (4.48 ppm), C-1 (103.40 ppm) and the $J_{H1,2} \sim 8$ Hz established a β configuration. A signal at 60.35 ppm in the [13]C spectrum correlated with a strong single resonance at 3.50 ppm in the proton spectrum, which is assigned to the $O$-methyl group. The connectivity from H-1 to H-5 is clearly established from the COSY spectrum. Resonance of C-4 at 82.80 ppm indicated that the $O$-methyl group is linked at this position, which otherwise would be in the range of 72 to 74 ppm. All of these assignments are in agreement with literature values reported for 4-$O$-methyl or other types of 4-substituted β-D-glucuronic acid [16].

A complete assignment of the NMR signals of the 6-linked β-D-galactose (B) is also summarized in Table 1.10. A β configuration at 1 position is evidenced by the signals at 4.64 ppm (H-1), 104.16 ppm (C-1) and $J_{H1,2} \sim 8$ Hz. The connectivity from H-1 and H-6 is established from the COSY. Signals at 3.78 ppm ([1]H) and 61.89 ppm ([13]C) are assigned to the H-6 of 2-linked D-galactose, although the rest of the resonances could not be assigned due to overlap with the signals derived from the 6-linked residue. Signals originating from 2,4-linked α-L-rhamnose (C) are assigned. Signals at 5.31 ppm (H-1) and 99.18 ppm (C-1) confirm the α configuration, while pyranosyl ring connectivity is established by the COSY. These assignments agreed well with literature values for 2,4-linked α-L-rhamnose. The complete assignments of signals from the 2-linked α-L-rhamnose were obtained by COSY analysis starting from H-6 (1.28 ppm) to H-5 (3.84 ppm), followed by H-4,3,2,1 [16].

A complete assignment of the signals that originated from 4-linked α-D-galacturonic acid is obtained by comparing the observed chemical shifts with literature values (Table 1.10). α Configuration is established by signals at 5.05 ppm (H-1) and 98.53 ppm (C-1). The correlations between H-1s (5.31 ppm and 5.05 ppm, respectively) and H-2s (4.12 ppm and 3.68 ppm, respectively) of the α-L-rhamnose and the α-D-galacturonic acid are not observed in the COSY spectrum due to overlap. However, this problem is overcome by comparing the chemical shifts with literature values and by the NOESY analysis (Figure 1.10 and Table 1.11), in which clear correlations are observed for those anomeric resonances.

By examining the anomeric region of the NOESY spectrum of WSCP-III (Figure 1.9), linkage sites and sequence of the identified residues can be

**TABLE 1.10.** $^1$H and $^{13}$C NMR Data and Assignments for Polysaccharide WSCP-III.

| Residue | | Chemical Shift (ppm) | | |
|---|---|---|---|---|
| | | Proton | | Carbon-13 |
| 4-O-Me-β-D-Glc pA-(1→ | H-1 | 4.48 | C-1 | 103.4 |
| | H-2 | 3.38 | C-2 | 73.98 |
| | H-3 | 3.57 | C-3 | 75.98 |
| **A** | H-4 | 3.33 | C-4 | 82.8 |
| | H-5 | 3.70 | C-5 | 77.29 |
| | | | C-6 | 176.22 |
| | H-4-O-Me | 3.50 | C-4-O-Me | 60.48 |
| →6)-β-D-Gal p-(1→ | H-1 | 4.64 | C-1 | 104.16 |
| | H-2 | 3.54 | C-2 | 72.44 |
| **B** | H-3 | 3.66 | C-3 | 73.53 |
| | H-4 | 3.91 | C-4 | 69.72 |
| | H-5 | 3.88 | C-5 | 74.64 |
| | H-6 | 3.98 | C-6 | 69.62 |
| | | 3.76 | | |
| ↓ | H-1 | 5.31 | C-1 | 99.16 |
| 4 | H-2 | 4.12 | C-2 | 77.29 |
| →2)-α-L-Rha p-(1→ | H-3 | 3.98 | C-3 | 69.62 |
| | H-4 | 3.68 | C-4 | 82.02 |
| **C** | H-5 | 4.10 | C-5 | 71.48 |
| | H-6 | 1.30 | C-6 | 17.75 |
| →2)-β-L-Rha p-(1→ | H-1 | 5.29 | C-1 | 99.18 |
| | H-2 | 4.12 | C-2 | 77.29 |
| **C′** | H-3 | 3.98 | C-3 | 69.62 |
| | H-4 | 3.68 | C-4 | 73.98 |
| | H-5 | 4.10 | C-5 | 69.62 |
| | H-6 | 1.28 | C-6 | 17.50 |
| →4)-α-D-Gal pA-(1→ | H-1 | 5.05 | C-1 | 98.53 |
| | H-2 | 3.95 | C-2 | 69.62 |
| **D** | H-3 | 4.12 | C-3 | 71.34 |
| | H-4 | 4.44 | C-4 | 77.92 |
| | H-5 | 4.62 | C-5 | 72.49 |
| | | | C-6 | 175.39 |

determined, as shown in Table 1.11. The anomeric resonance (4.48 ppm) of the terminal nonreducing 4-O-Me-β-D-glcpA (residue **A** in Tables 1.10 and 1.11) correlated to the H-6 (3.98 ppm) of the 6-linked β-D-galp (residue **B** in Tables 1.10 and 1.11). NOE correlations are also observed between the H-1 (4.64 ppm) of residue **B** and the H-4 (3.68 ppm) of residue **C** and H-6 (3.98 ppm) of residue **B**, respectively. This observation suggests that in some cases, two or

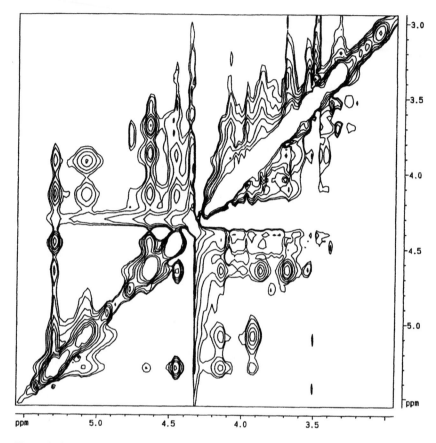

**Figure 1.10** NOESY spectrum of WSCP-III recorded at 65°C in $D_2O$. Reprinted from Reference [16] with permission from Elsevier Science.

more consecutive **B** residues are present. H-1 of **C** or **C′** (5.29 and 5.31 ppm) simultaneously correlated with H-2 (4.12 ppm) of **C** or **C′** and H-4 (4.44 ppm) of residue **D**, the 4-linked α-D-galpA.

In order to elucidate and confirm the structure obtained from NMR analysis, WSCP-III is partially hydrolyzed in 0.4 M trifluoroacetic acid at 100°C for 2.5 hr [16]. Three fractions (F-I, F-II and F-III, respectively) were obtained by ion exchange chromatography (DEAE-Sephadex A-25 column, 1.6 × 20 cm, 25 mM NaOAc buffer at pH 5) eluted with increased ion strength [16]. The three oligosaccharides are reduced to alditols by borohydride to simplify their NMR spectra. In the region of anomeric protons (4.4 to 4.5) of the $^1H$ NMR spectra, two doublets were observed for F-I, while only one doublet was found for both F-II and F-III. The presence of two doublets suggests a trisaccharide for F-I; using the same principle, a single doublet indicates that F-II and F-III are disaccharides because the reducing terminal C1s were reduced to alditols by

**TABLE 1.11. Observed NOE Contacts from Anomeric Protons of Polysaccharide WSCP-III.**

| Anomeric Proton (ppm) | | NOE Contacts to (ppm) | | | |
|---|---|---|---|---|---|
| 4.48 | A | 4-$O$-Me-β-D-Glc pA-(1→ | 3.98 | B | H-6 |
| 4.64 | B | →6)-β-D-Gal p-(1→ | 3.68 | C | H-4 |
| 5.31 | C | R$^a$ | 4.12 | C | H-2 |
| | | ↓ | | | |
| | | 4 | 4.44 | D | H-4 |
| | | →2)-α-L-Rha p-(1→ | | | |
| 5.05 | D | →4)-α-D-Gal pA-(1→ | 4.12 | C | H-2 |

$^a$ $R$ = H, the structure is C', without the 4-linkage. $R$ = B, the structure is C, with the 4-linkage.
Reprinted from Reference [16] with permission from Elsevier Science.

borohydride [16]. Two-dimensional NMR spectroscopy was carried out to establish the connectivities and linkage sites of those oligosaccharides. As shown in Table 1.12, full assignments of all resonances are obtained by comparing chemical shifts against literature values with the assistance of homonuclear shift-correlated spectroscopy. The COSY experiments established the scalar coupling connectivities of the oligosaccharides that supported the complete assignment of the resonances (Table 1.12).

F-I is established as a disaccharide alditol that contained terminal nonreducing 4-$O$-Me-β-D-GlcpA, 6-linked β-D-Galp and 4-linked rhamnitol, as shown in Structure 1.

$$4\text{-}O\text{-Me-}β\text{-D-GlcpA-}(1 \rightarrow 6)\text{-}β\text{-D-Galp-}(1 \rightarrow 4)\text{-L-rhamnitol}$$

Structure 1 (F-I)

The β configurations were apparent for the 4-$O$-Me-D-GlcpA and the 6-linked D-Galp by $J_{H1,2}$ (~8 Hz) as well as by the chemical shifts of the anomeric protons and carbons (Table 1.12). Rhamnitol residue is evidenced by the presence of 1.29 ppm ($^1$H) and 17.85 ppm ($^{13}$C), which arose from the CH$_3$ group in deoxy sugars. The sequence and site of the linkage of F-I are established by a differential NOE experiment, and the result is shown in Table 1.13.

F-II and F-III are monosaccharide alditols that contained 4-$O$-Me-β-D-GlcpA as the terminal nonreducing groups. The two alditols were identified as D-galactitols. The differential NOE experiment suggested that the 4-$O$-Me-β-D-GlcpA linked to the 6 position in F-II and to the 2 position in F-III (Table 1.13). The two structures are shown below.

$$4\text{-}O\text{-Me-}β\text{-D-GlcpA-}(1 \rightarrow 6)\text{-D-Galactitol}$$

Structure 2 (F-II)

$$4\text{-}O\text{-Me-}β\text{-D-GlcpA-}(1 \rightarrow 2)\text{-D-Galactitol}$$

Structure 3 (F-III)

TABLE 1.12. $^1H$ and $^{13}C$ NMR Data and Assignment of Three Oligosaccharide Alditols Derived from Polysaccharide WSCP-III.

| Residue | $^1H$ | F-I ($J_{1,2}$) | F-II ($J_{1,2}$) | F-III ($J_{1,2}$) | Chemical Shift ($\delta$/ppm) $^{13}C$ | F-I | F-II | F-III |
|---|---|---|---|---|---|---|---|---|
| 4-O-Me-β-D-Glc pA-(1→ | H1 | 4.49 (8 Hz) | 4.48 (8 Hz) | 4.46 (8 Hz) | C-1 | 103.60 | 103.60 | 103.00 |
| | H2 | 3.38 | 3.39 | 3.38 | C-2 | 73.98 | 73.80 | 73.80 |
| | H3 | 3.57 | 3.56 | 3.57 | C-3 | 76.00 | 76.00 | 76.00 |
| | H4 | 3.34 | 3.34 | 3.33 | C-4 | 82.69 | 83.00 | 83.00 |
| | H5 | 3.70 | 3.71 | 3.72 | C-5 | 77.29 | 77.35 | 77.30 |
| | | | | | C-6 | [176.22] | [176.22] | [176.22]ᵃ |
| | H-4-O-Me | 3.50 | | | C-4-O-Me | 60.54 | 60.52 | 60.35 |
| →6)-β-D-Gal p-(1→(F-I) | H-1 | 4.64 (8 Hz) | 3.70 | 3.74 | C-1 | 104.16 | 64.16 | 63.47 |
| →6)-D-Galactitol (F-II) | H-2 | 3.54 | 3.95 | 4.00 | C-2 | 72.44 | 72.50 | 72.40ᵇ |
| →2)-D-Galactitol (F-III)ᵇ | H-3 | 3.66 | 3.96 | 3.70 | C-3 | 73.53 | 74.04 | 71.71 |
| | H-4 | 3.91 | 3.72 | 3.68 | C-4 | 69.72 | 71.00 | 71.90 |
| | H-5 | 3.88 | 4.15 | 3.84 | C-5 | 74.64 | 74.00 | 72.00 |
| | H-6 | 3.98 | 3.98 | 3.65 | C-6 | 69.62 | 69.50 | 63.47 |
| | | 3.76 | 3.68 | | | | | |
| →4)-L-Rhamnitol | H-1 | 3.73 | | | C-1 | 64.17 | | |
| | H-2 | 3.94 | | | C-2 | 70.71 | | |
| | H-3 | 3.98 | | | C-3 | 69.62 | | |
| | H-4 | 3.84 | | | C-4 | 75.95 | | |
| | H-5 | 4.14 | | | C-5 | 69.90 | | |
| | H-6 | 1.29 | | | C-6 | 17.50 | | |

ᵃResonance was not determined because the $^{13}C$ resonances were obtained from proton-detected spectrum; data presented were from the polysaccharide.
ᵇTentative assignment.
Reprinted from Reference [16] with permission from Elsevier Science.

TABLE 1.13. Different NOE Inter-Residue Connectivities
of Oligosaccharide Alditols Derived from Polysaccharide WSCP-III.

| | | Anomeric Proton (ppm) | NOE Inter-Residue Connectivity (ppm) | | | |
|---|---|---|---|---|---|---|
| F-I | | | | | | |
| 4-O-Me-β-D-Glc pA-(1→ | a | 4.49 | 3.98 | H-6 | of | b |
| →6)-β-D-Gal p-(1→ | b | 4.46 | 3.84 | H-4 | of | c |
| →4)-L-rhamnitol | c | | | | | |
| F-II | | | | | | |
| 4-O-Me-β-D-Glc pA-(1→ | a′ | 4.48 | 3.98 | H-6 | of | b′ |
| →6)-D-galactitol | b′ | | | | | |
| F-III | | | | | | |
| 4-O-Me-β-D-Glc pA-(1→ | a″ | 4.46 | 4.00 | H-2 | of | b″ |
| →6)-D-galactitol | b″ | | | | | |

Reprinted from Reference [16] with permission from Elsevier Science.

The structural features and linkage patterns identified for the three oligosaccharides are in agreement with the result from two-dimensional NMR spectroscopy of the polysaccharide.

In summary, WSCP-III is identified as a pectic polysaccharide composed of a disaccharide backbone repeating unit:

$$\rightarrow 2)\text{-}\alpha\text{-L-Rhamp-}(1\rightarrow 4)\text{-}\alpha\text{-D-GalpA-}(1\rightarrow$$

Oligosaccharide side chains are attached to the 4 position of the 2-linked α-L-rhamnose residue. The ratio of the 4-substituted and unsubstituted 2-linked α-L-rhamnose is 2:1. The side chains are composed of a terminal nonreducing end 4-O-Me-β-D-GlcpA that is attached to the 4 position of the 2-linked α-L-Rhamp in the backbone chain, mainly through 6-linked (a small portion through 2-linked) β-D-Galp. Rhamnogalacturonan structure is common in the plant kingdom. The identified →2)-α-L-Rhap-(1→4)-α-D-GalpA-(1→ backbone chain is typical of pectic polysaccharides [18].

## 4. PHYSICAL PROPERTIES OF YMG

### 4.1. FUNCTIONAL PROPERTIES OF YMG

Water-holding capacity, emulsion stability and foam stability are important functional properties of commercial gums. However, there is only limited information available in the literature concerning these functional properties of

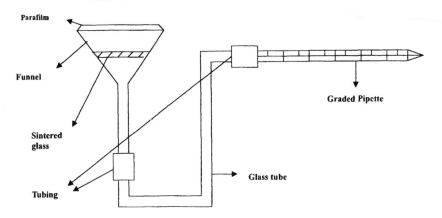

Parafilm

Funnel

Sintered
glass

Tubing

Graded Pipette

Glass tube

**Figure 1.11** A scheme of a device for determination of the water holding capacity of hydrocolloid gums.

YMG, partially due to lack of reliable and accurate methods. Cui and Eskin reported that the water-holding-capacity (WHC) of YMG was 3446 g $H_2O$/100 g of solid compared to 20,314 g $H_2O$/100 g of solid for xanthan gum and 2961 g $H_2O$/100 g of solid for guar gum [19]. The measurement device is designed according to the principles of the Baumann's capillary apparatus and is easy to set up in any laboratory with regular glassware, as shown in Figure 1.11. The operating procedure is as follows: (1) fill the pipette and funnel below the sintered glass with distilled water without forming any air bubbles (a syringe is useful in removing air bubbles); (2) place a piece of filter paper (the size of the sintered glass) on top of the sintered glass and allow to equilibrate—the water level at the graded pipette end is adjusted to zero by adding or absorbing additional water to or from the funnel; (3) accurately weigh 10 to 30 milligrams of powdered gum samples (40–60 mesh) and place the sample on top of the filter paper, sealing the top of the funnel with parafilm; (4) take a reading of the water absorbed by the sample immediately after loading the sample, and take further readings every 2 hr; (5) use a blank to correct the effects of evaporation. The absorption rate can be obtained by plotting the absorbed water against time. The amount of water absorbed at the equilibrium is taken to calculate the WHC. The results of selected gums are presented in Table 1.14.

YMG contained surface active component as it can reduce the surface tension of water, as shown in Figure 1.12. Increasing the gum concentration to 0.05% substantially reduced surface tension. Further additions of mucilage only slightly decreased the surface tension. WS exhibited the greatest reduction in surface tension compared to the crude mucilage (CM) and the water- insoluble (WI) fraction. The surface and interfacial activities of some plant hydrocolloids

TABLE 1.14. **Comparison of Water-Holding Capacity (WHC) of Gums by the Modified and Baumann Methods.**

| | Methods | | | |
| | Modified | | Baumann | |
| Sample | WHC (gH$_2$O/100 g solid) | Rel. Error (%) | WHC (gH$_2$O/100 g solid) | Rel. Error (%) |
|---|---|---|---|---|
| Xanthan gum | 20134 ± 883[a] | 4 | 15850 ± 7460[b] | 47 |
| Guar gum | 2961 ± 200 | 10 | 2480 ± 400[c] | 19 |
| Yellow mustard gum | 3446 ± 187 | 5 | | |

[a] $n = 3$, means ± s.d.
[b] $n = 4$, literature value [19].
[c] $n = 3$, literature value [19].

(guar and locust bean gums, etc.) were recently ascribed to the presence of residual surface active constituents or impurities in them. The protein present in CM and its fractions could contribute to the surface activity of YMG; however, the WS fraction, which had the lowest protein content, was the most surface active fraction. The lowest interfacial tension of vegetable oil and water was also reported for YMG compared to eight commercial gums [3]. Application of the surface/interfacial properties of gums is reflected by their ability to stabilize emulsions and foams. The emulsion capacity and stability of YMG are determined according to a method described by Yasumatsu et al. for protein [20], with modification. Gum sample (0.50 g) is suspended in 40 mL of distilled water before mixing with 40 mL of vegetable oil. The water/oil mixture is emulsified using a polytron at 10,000 rpm for 60 s, and then it is centrifuged at 1300 g for 5 min. Emulsion capacity is calculated as follows:

$$\frac{\text{height of emulsion layer}}{\text{total height of fluid}} \times 100\%$$

Emulsion stability is determined by heating the emulsion at 80°C for 30 min, cooling with tap water for 15 min, then centrifuging at 1300 g for 5 min. Emulsion stability is calculated as follows:

$$\frac{\text{height of remaining emulsion layer}}{\text{total height of fluid}} \times 100\%$$

The emulsion capacity and stability of yellow mustard mucilage and its fractions were compared to commercial gums as shown in Figure 1.13. Before dialysis, the CM exhibited the highest emulsion capacity and stability as compared

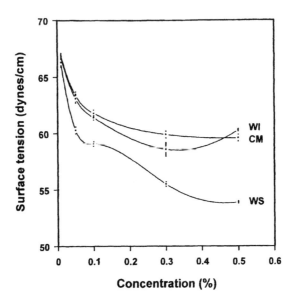

**Figure 1.12** Reduction of surface tension of water by CM, WS and WI at various concentrations; the corrected surface tension of distilled water was $69.0 \pm 0.7$ dyne/cm at $23.0 \pm 0.5°C$ (CM = crude mucilage; WS = water-soluble fraction; WI = water-insoluble fraction). Reprinted from Reference [9] with permission from Elsevier Science.

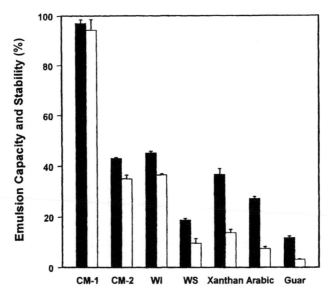

**Figure 1.13** Emulsion capacity (filled) and stability (open) of yellow mustard mucilage fractions and other commercial gums; CM-1 and CM-2 are crude mucilage samples before and after dialysis. Reprinted from Reference [9] with permission from Elsevier Science.

to other gum fractions or commercial gums. Although dialysis reduced the emulsion capacity and stability of YMG substantially, it still exhibited higher emulsion capacity and stability compared to the commercial gums examined. Earlier research by Weber and coworkers suggested that emulsion stability increased as the concentration of YMG increased; and higher gum concentration tended to give a smaller droplet size [3]. Blends of YMG with other gums (at the ratio of maximum synergism, as described in section 5) demonstrated above-average emulsion stability, as shown in Table 1.15.

Foaming capacity and stability of YMG and commercial gums are determined by dissolving 0.3% of gums in 0.1% ovalbumin; foaming capacity and

**TABLE 1.15. Emulsification of 40% Vegetable Oil and Water Produced by Hydrocolloids.**

| | Viscosity @ 20 rpm cps | Droplet Range μ | Avg. Droplet Size μ | Emulsion Stability |
|---|---|---|---|---|
| 0.25% Mustard mucilage | 660 | 10–75 | 30 | 52.00% |
| 0.50% Mustard mucilage | 1810 | 5–50 | 20 | 82.60% |
| 0.75% Mustard mucilage | 3375 | 2–50 | 15 | 99.50% |
| 1.00% Mustard mucilage | 5800 | 2–35 | 8 | 100.00% |
| 0.5% Locust bean | 1615 | 15–90 | 45 | 63.50% |
| 0.5% (0.8 Locust bean/0.2) | 2463 | 5–65 | 35 | 85.00% |
| 0.5% Guar | 3600 | 20–150 | 90 | 52.00% |
| 0.5% (0.9 Guar/0.1 Muc.) | 3900 | 10–80 | 50 | 81.00% |
| 0.5% CMC | 135 | 10–200 | 40 | 65.00% |
| 0.5% (0.6 CMC/0.4 Muc.) | 1725 | 5–130 | 20 | 99.50% |
| 0.5% Xanthan | 3000 | 5–125 | 8 | 100.00% |
| 0.5% (0.8 Xan./0.2 Muc.) | 2350 | 2–80 | 10 | 100.00% |
| 0.5% Tragacanth | 1110 | 10–135 | 45 | 74.50% |
| 0.5% (0.7 Trag./0.3 Muc.) | 1515 | 5–80 | 15 | 89.00% |
| 0.5% Propylene glycol alginate | 160 | 2–60 | 5 | 100.00% |
| 0.5% (0.5 PGA/0.5 Muc.) | 970 | 2–70 | 15 | 100.00% |

Reprinted from Reference [3] with permission.

TABLE 1.16. Effect of Yellow Mustard Gum on Foaming Capacity and Stability of 0.1% Bovine Serum Albumin Solutions.

| Time (h) | Foam Volume (mL) | | | | | | |
|---|---|---|---|---|---|---|---|
| | CM[a] | CM[b] | WS | WI | Xanthan | Arabic | Guar |
| 0.0 | 23.5 ± 0.5[c] | 38.3 ± 0.6 | 52.3 ± 1.5 | 39.0 ± 1.0 | 36.5 ± 0.9 | 41.8 ± 1.0 | 12.8 ± 0.3 |
| 0.5 | 19.2 ± 1.0 | 16.5 ± 1.3 | 29.2 ± 0.7 | 16.7 ± 0.8 | 36.5 ± 1.0 | 36.3 ± 1.2 | 9.3 ± 0.3 |
| 1 | 18.3 ± 0.8 | 13.7 ± 0.3 | 20.3 ± 0.4 | 15.0 ± 1.3 | 36.2 ± 0.5 | 32.8 ± 1.3 | 8.2 ± 0.0 |
| 3 | 15.9 ± 0.1 | 6.3 ± 1.0 | 0.0 ± 0.0[d] | 0.0 ± 0.0 | 18.7 ± 1.5 | 20.6 ± 0.6 | 7.3 ± 0.6 |
| 5 | 14.8 ± 0.3 | 1.0 ± 0.0 | 0.0 ± 0.0 | 0.0 ± 0.0 | 17.5 ± 1.3 | 8.0 ± 0.9 | 6.6 ± 0.5 |
| 6 | 14.7 ± 0.3 | 0.0 ± 0.0 | 0.0 ± 0.0 | 0.0 ± 0.0 | 17.2 ± 1.0 | 6.6 ± 1.3 | 5.9 ± 0.1 |
| 23 | 13.3 ± 0.3 | 0.0 ± 0.0 | 0.0 ± 0.0 | 0.0 ± 0.0 | 13.2 ± 1.3 | 0.5 ± 0.5 | 4.0 ± 0.0 |

[a]CM before dialysis.
[b]CM after dialysis.
[c]$n = 3$, mean ± SD.
[d]Diminished.
CM = crude mucilage; WS = water-soluble fraction; WI = water-insoluble fraction.
Reprinted from Reference [9] with permission from Elsevier Science.

stability of YMG and its fractions and commercial gums are summarized in Table 1.16. Prior to dialysis, YMG exhibited the highest foaming stability among the yellow mustard mucilage fractions. Following dialysis, the foaming capacity of YMG increased, but its foam is less stable. Similar trends are also observed for the WS and WI fractions.

## 4.2. RHEOLOGY: FLOW BEHAVIORS

The flow profiles of the water-soluble YMG (WS) and its fractions are shown in Figure 1.14 and 1.15, respectively. Typical shear thinning flow behaviors are observed for WS, WSCP and WSCS at all concentrations examined (0.3% to 2.0%, w/w). A Newtonian plateau is not evident at concentrations between 0.5% to 2.0% for the three samples over a wide shear rate range (0.1 to 1162 s$^{-1}$). However, Newtonian plateau (upper) is observed at a concentration of 0.3%. The zero-shear-rate viscosity and the shear rate value ($\dot{\gamma}$) at which the onset shear-thinning flow behavior occurs are shown in Table 1.17. The zero-shear-rate viscosity is found to be the highest for WS, followed by WSCP and WSCS, while the $\dot{\gamma}$ value at which the onset shear-thinning flow behavior occurs is the highest for WSCS, lowest for WSCP and in between for WS. WSCP and WSCS appeared to contribute to the viscoelastic character of WS solutions. The shear thinning flow behaviors, i.e., the decrease in viscosity as shear rate increases, observed for yellow mustard gum and its fractions resembled those of xanthan gum dispersions at all concentrations examined.

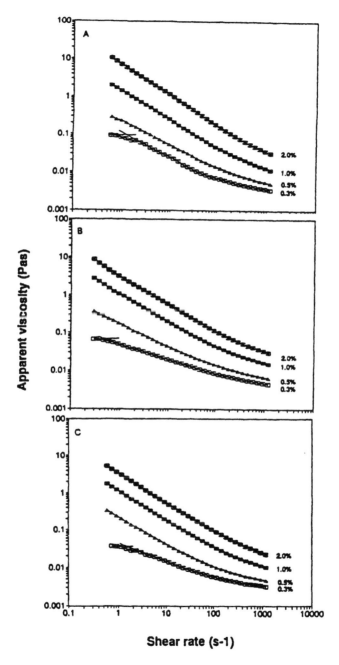

**Figure 1.14** Steady shear flow curves of the yellow mustard mucilage water-soluble fraction (WS, A) and its subfractions: the CTAB-precipitated fraction (WSCP, B) and the CTAB-soluble fraction (WSCS, C) at concentrations between 0.3 and 2.0%, 22.0 ± 0.1°C. Reprinted from Reference [14] with permission from Elsevier Science.

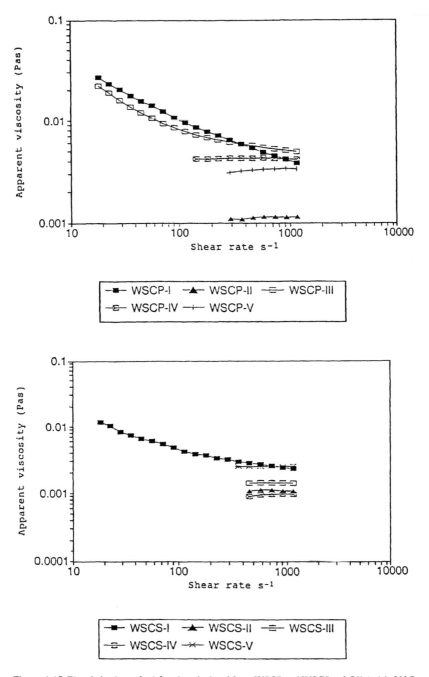

**Figure 1.15** Flow behaviors of subfractions isolated from WSCP and WSCS at 0.5% (w/v), 22°C.

TABLE 1.17. **A Comparison of Zero-Shear-Rate Viscosity ($\eta_0$) and Shear-Rate ($\dot{\gamma}$) Value at which Onset Shear Thinning Occurred for the Water-Soluble Fraction (WS) and Its Subfractions: CTAB-Precipited Fraction (WSCP) and CTAB-Soluble Fraction (WSCS) at 0.3% (22°C).**

|      | $\eta_0$ (MPa s) | $\dot{\gamma}$ Value (s$^{-1}$) |
|------|------------------|---------------------------------|
| WS   | 94.4             | 1.16                            |
| WSCP | 68.2             | 0.46                            |
| WSCS | 36.4             | 1.47                            |

Reprinted from Reference [14] with permission from Elsevier Science.

By applying the power law model [21], the consistency index ($K$) and the flow index ($n$), defined from the equation $\eta = K\dot{\gamma}^{n-1}$, are obtained (Table 1.18). The consistency index "$K$" is proportional to the measured viscosity, while the flow index "$n$" measures the pseudoplasticity of the system. As concentration increases, the "$K$" increases while $n$ decreases. The increases in "$K$" with increasing concentration suggest that a more viscous system is obtained at higher concentrations. On the other hand, the decrease in "$n$" value with increasing concentration implies a more pronounced shear thinning flow behavior. WS exhibited the highest "$K$" value and the lowest "$n$" value compared to those of WSCP and WSCS, suggesting the presence of possible polysaccharide-polysaccharide interactions between some of the components present in WSCP and WSCS. This phenomenon is particularly significant at 2.0% (w/w) polymer solution.

The flow behavior of subfractions from WSCP and WSCS is presented in Figure 1.15. In the WSCP series, WSCP-I and WSCP-III exhibited shear thinning flow behavior at 0.5% (w/w), while WSCP-II, IV and V are Newtonian flows [Figure 1.15(a)]. In the WSCS series, only WSCS-I exhibited shear thinning flow behavior [Figure 1.15(b)]. It is concluded that of the ten

TABLE 1.18. **$n$ And $k$ values[a] of the Water-Soluble Fraction (WS) and Its Subfractions: CTAB-Precipitated Fraction (WSCP) and CTAB-Soluble Fraction (WSCS) at Different Concentrations (22°C).**

| Concentration (%) | WSCP | | WSCS | | WS | |
|-------------------|-------|-----------|-------|-----------|-------|-----------|
|                   | $n$   | $K$ (Pas) | $n$   | $K$ (Pas) | $n$   | $K$ (Pas) |
| 0.3               | 0.668 | 0.041     | 0.641 | 0.033     | 0.537 | 0.070     |
| 0.5               | 0.557 | 0.119     | 0.473 | 0.138     | 0.468 | 0.173     |
| 1.0               | 0.407 | 0.740     | 0.331 | 0.845     | 0.321 | 1.120     |
| 2.0               | 0.340 | 2.506     | 0.285 | 2.764     | 0.224 | 6.440     |

[a]Parameters $n$ and $K$ were calculated using the power-law model: $\eta = k\gamma^{n-1}$.
Reprinted from Reference [14] with permission from Elsevier Science.

subfractions isolated from water-soluble YMG, three fractions (WSCP-I, WSCP-III and WSCS-I) exhibited shear thinning flow behavior originally observed for YMG.

## 4.3. RHEOLOGY: VISCOELASTIC PROPERTIES

The small strain oscillatory rheological tests of YMG and its fractions revealed that YMG solutions/dispersions exhibit weak gel structures (Figure 1.16). In a weak gel structure, such as xanthan gum dispersions, the storage modulus $G'$ is greater than the loss modulus $G''$, and both moduli are only slightly frequency dependent. Contrary to viscoelastic fluid, such as guar gum solution, its mechanical spectrum is characterized by $G' \propto \omega^2$ and $G'' \propto \omega^1$, and in most situations, $G'' > G'$ at low frequencies, and the reverse occurs at high frequencies. The rheological responses of different systems can also be described by the phase angle changes vs. frequency. The tangent of phase angle $\delta$, defined as the ratio of $G''/G'$, expresses the relative contributions of the viscous and elastic components to the visoelastic properties of a system. The slight increases of $G'$ and $G''$ vs. frequency resulted in relatively constant phase angle

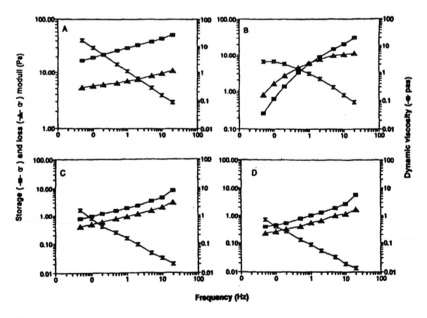

**Figure 1.16** Frequency dependence of storage ($G'$) and loss ($G''$) moduli and viscosity ($\eta'$) of (A) xanthan gum, (B) guar gum, (C) CM and (D) WS for 1.0% (w/w) solutions/dispersions (CM = crude mucilage; WS = water-soluble fraction). Reprinted from Reference [9] with permission from Elsevier Science.

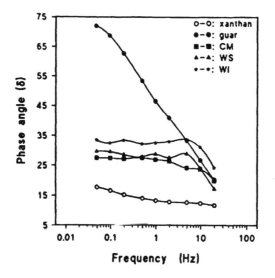

**Figure 1.17** Phase angle value profiles as a function of frequency of yellow mustard mucilage and commercial gum for 1% (w/w) solutions/dispersions (CM = crude mucilage; WS = water-soluble fraction; WI = water-insoluble fraction). Reprinted from Reference [9] with permission from Elsevier Science.

values for xanthan and yellow mustard gum solutions over the entire frequency range examined (Figure 1.17). In contrast, the phase angle values of viscoelastic fluid, such as guar gum solution, are highly dependent on frequency. WSCP and WSCS exhibited similar dynamic rheological patterns to that of WS, while the maximum frequency dependence is approximately $G' \propto \omega^{0.15}$ and $G'' \propto \omega^{0.25}$ for WSCP, $G' \propto \omega^{0.23}$ and $G'' \propto \omega^{0.22}$ for WSCS as compared to $G' \propto \omega^{0.16}$ and $G'' \propto \omega^{0.21}$ for WS. Figure 1.18 shows the changes of $G'$ as a function of polymer concentration. Compared to WSCP, whose $G'$ coincides with that of WS, the $G'$ of WSCS appeared similar to that of WS only at low concentrations and departed from that of WS at higher concentrations (e.g., 2.0%). The non-linear increase of $G'$ of WSCP and WS may reflect that a more ordered network structure develops in WS and WSCP solutions at higher concentrations.

The relationship between the apparent viscosity ($\eta$) and complex viscosity ($\eta^*$, defined as $\eta^* = [G'^2 + G''^2]^{0.5}/\omega$) can be used diagnostically to distinguish a normal viscoelastic fluid from weak gels. For a weak gel system, the complex viscosity $\eta^*(f)$ is substantially higher than the apparent viscosity $\eta(\dot{\gamma})$ at equivalent values of frequency and shear rate. In contrast, the two viscosities coincide in a normal viscoelastic fluid; this is known as the Cox-Merz rule [21]. The apparent viscosity $\eta(\dot{\gamma})$ and complex viscosity $\eta^*(f)$ of YMG samples are presented in Figure 1.19. The complex viscosity $\eta^*(f)$ of WS is substantially higher than the apparent viscosity $\eta(\dot{\gamma})$ at equivalent values of frequency and

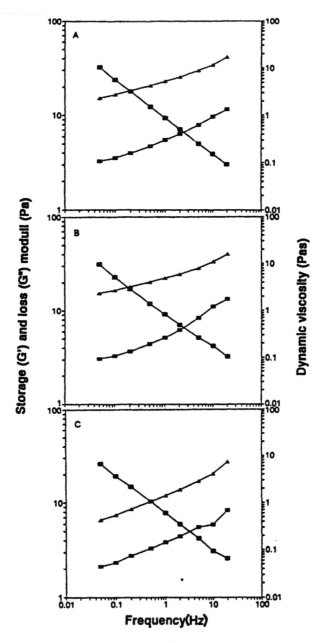

**Figure 1.18** Frequency dependence of storage ($G'$, ▲) and loss ($G''$, ■) moduli and dynamic viscosity ($\eta'$, ❑) of yellow mustard mucilage water-soluble fraction (WS, A) and its subfractions: CTAB-precipitated fraction (WSCP, B) and CTAB-soluble fraction (WSCS, C) at concentration 2.0%, 22.0 ± 0.1°C. Reprinted from Reference [14] with permission from Elsevier Science.

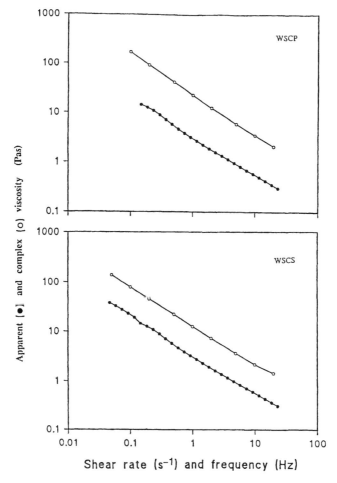

**Figure 1.19** Cox-Merz plot for 2% solutions or dispersions of the water-soluble fraction of yellow mustard mucilage: CTAB-precipitated fraction (WSCP, ○) and CTAB-soluble fraction (WSCS, ●) at 22.0°C. Reprinted from Reference [14] with permission from Elsevier Science.

shear rate, thus exhibiting the typical behavior of a weak gel (Figure 1.19). The extent of $\eta^*(f)$ over $\eta(\dot{\gamma})$ for WSCP is greater than that for WS, indicating a more ordered weak gel structure or more elastic character for the WSCP fraction. The Cox-Merz rule was also tested for the WSCS fraction over the whole range of frequencies and shear rates investigated; again, $\eta^*(f)$ and $\eta(\dot{\gamma})$ were not superimposible as shown in Figure 1.19. The degree of departure of $\eta^*(f)$ from $\eta(\dot{\gamma})$, however, was much smaller for WSCS compared to WSCP. Such a departure from the Cox-Merz superimposability has been attributed to the making and breaking of non-covalent (hydrogen) bonds [22]. The rheological properties discussed above suggested that both WSCP and WSCS contribute to the rheological properties of WS. Being the major fraction, WSCP contributes more to shear thinning and weak gel properties of WS, while WSCS contributes more to the viscous properties of WS solutions.

## 4.4. EFFECT OF pH, TEMPERATURE, SALT AND SUGAR ON APPARENT VISCOSITY OF YMG

The effects of pH on rheological properties of YMG solutions (0.5% w/w) are shown in Figure 1.20. Viscosity at a shear rate of 93.32 s$^{-1}$, approximating

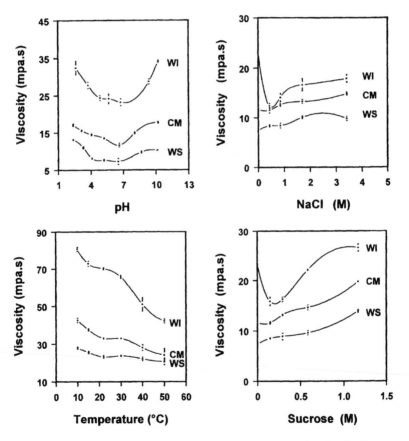

**Figure 1.20** Effect of pH, temperature, salt and sugar on the apparent viscosity of yellow mustard mucilage fractions at shear rate 92.32 s$^{-1}$. (concentration for temperature effect: 1.0% w/w; concentrations for pH, salt and sugar effect: 0.5% w/w; pH for temperature, salt and sugar effects: 6.3). Reprinted from Reference [9] with permission from Elsevier Science.

mouthfeel conditions, increased at lower and higher pH regions, which is in agreement with the earlier work of Weber and co-workers [3]. The increase in viscosity at low and high pH regions suggested that acid or alkaline environments altered the conformation of the polysaccharides and most likely affected the intermolecular interactions due to modification of electrostatic effects. Furthermore, the viscosity increase of WI in the alkaline region could be due to a more effective dispersion or solubilization of the insoluble gum fraction under such conditions. The extent of the influence of pH on the apparent viscosity of gum solutions (suspensions) is in the order of WI > CM > WS.

Similar trends in the change of viscosity are observed for YMG and its fractions (at 0.5% w/w) when salt and sugar are added. Solutions of YMG and WS generally exhibited higher viscosity values with increasing additive concentrations. In contrast, WI dispersions exhibited an initial reduction in viscosity at low solute concentrations (<0.5 M NaCl, <0.25 M sucrose). At much higher solute concentrations, the rheological responses of WI are similar to those of YMG and WS. A study by Anguilar and Ziegler showed that temperature and electrolytes had a negative effect on the viscosity of aqueous dispersions of mustard seed mucilage [23].

The effects of temperature on the apparent viscosity of YMG solutions/dispersions are also presented in Figure 1.20. An increase in temperature resulted in a continuous reduction in viscosity. The reduction in viscosity is more pronounced for WI than the other two fractions. It suggests the presence of extended interparticle associations for this material at low temperatures. The results of this study demonstrated the similarities of interfacial activities and rheological properties of YMG and xanthan gum solutions/dispersions, which suggest that yellow mustard gum could be used as a substitute for xanthan gum. The effect of temperature on the viscosities of WSCP and WSCS obeyed the expected trend of decreasing viscosity with increasing temperature (Figure 1.21). The influence of pH on the viscosities of WSCP and WSCS is also in agreement with previous reports [9] that the lowest viscosity was observed between pH 3 to 7 and the higher viscosity at the low pH region. The substantial increase in viscosity at the low pH region could be attributed to the reduction of repulsion forces between polymer chains that allows interchain associations between polymer molecules, thereby increasing viscosity. The addition of sucrose resulted in enhanced viscosity for WSCP and WSCS as shown in Figure 1.21. It appeared that the initial addition of sucrose brought about a faster increase in viscosity for WSCP with a possible turning point at 0.4 M sugar concentration; above this concentration, the rate of increase in viscosity became smaller. In contrast to WSCP, sucrose caused an almost linear increase in viscosity for WSCS. The effect of sucrose on the apparent viscosity of polysaccharide solutions can be attributed to a concentrating effect and improved polymer-solvent interactions. The responses of the apparent viscosities of WSCP and WSCS to the addition of salt are different, as shown in Figure 1.21. The initial addition of NaCl resulted

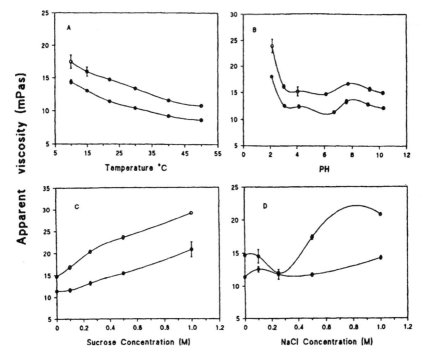

**Figure 1.21** Effect of temperature (A), pH (B), sucrose (C) and salt (D) concentrations on the apparent vicosity of water-soluble yellow mustard mucilage subfractions: CTAB-precipitated fraction (WSCP, O-O) and CTAB-soluble fraction (WSCS, ●-●) at 0.5% (w/w) polymer concentration. 22.0°C, shear rate 92.32 s$^{-1}$. Reprinted from Reference [14] with permission from Elsevier Science.

in a reduction in viscosity for the WSCP solutions up to 0.25 M salt concentration. Further addition of NaCl resulted in a rapid recovery of viscosity, and the increase of viscosity is more pronounced at higher concentrations of salt (~1.0 M). The initial reduction in viscosity with the addition of NaCl is attributed to the progressive suppression of intramolecular charge-charge repulsion and the consequent contraction of the polysaccharide molecules. As the salt concentration increased beyond a certain point, intermolecular charge-charge repulsions were suppressed and intermolecular associations could occur resulting in the recovery of viscosity. Further increases in viscosity at higher salt concentrations could also be due to a concentrating effect in addition to charge suppression. In contrast to WSCP, WSCS exhibited fairly stable viscosity upon the addition of salt. The initial addition of salt even slightly increased the viscosity. This is possibly due to the nonreducing end glucuronic acid that is attached as side residue to the backbone chain. The initial addition of NaCl would suppress the charge-induced intermolecular repulsions, thereby enhancing the intermolecular interactions between the polymer chains. Nevertheless, the overall response of WSCS solutions to added NaCl is minor compared to that of WSCP, and this may also reflect the relatively fewer acidic groups present in WSCS.

## 5. INTERACTION BETWEEN YMG AND GALACTOMANNANS

### 5.1. SYNERGISTIC PHENOMENA

The first synergism between YMG and other polysaccharides was observed by Weber and coworkers [3]. They observed that guar gum and YMG reached the maximum synergistic effect at 0.9% guar:0.1% YMG with a viscosity of 20,400 cps (2.3 times the theoretical additive value). The maximum viscosity of YMG and locust bean gum, obtained at 0.8% locust bean gum:0.2% YMG, was 10,850 cps (14.5 times the theoretical value). A similar synergistic effect was also observed between YMG and carboxymethylcellulose (CMC) at 0.6% of CMC:0.4% YMG. In addition to the gums described above, smaller interactions (1–2 times the theoretical value) were noted between YMG and tragacanth, furcellaran, xanthan and propylene glycol alginate, but no interactions were found with agar, gum arabic, carrageenans, gelatin, karaya, pectin and ghatti [3].

In order to understand the mechanisms of the synergism between YMG and other polysaccharides, Cui et al. examine the interactions between YMG and a series of galactomannans, including locust bean gum, guar gum and fenugreek gum, using dynamic oscillatory rheological measurements [24].

Galactomannans have been known to interact with a number of other polysaccharides, such as xanthan gum, carageenan and alginates, resulting in increased viscosity and/or inducing/improving gelling properties. When yellow mustard gum is mixed with galactomannans, the viscosity of the mixed systems increases significantly, accompanied by changes in viscoelastic characteristics [24]. The mechanical spectrum of galactomannans represents a typical viscoelastic fluid where the storage modulus $(G')$ is lower than the loss modulus $(G'')$ at lower frequencies, with the reverse observed at higher frequencies. The mechanical spectrum of YMG is characteristic of a weak gel in which the storage modulus $(G')$ predominates over the loss modulus $(G'')$ over the entire frequency range examined [9]. YMG and LBG blends are prepared by mixing the two gums in the same concentration according to designed proportions, heating at 80°C for at least 30 min before rheological measurements are taken. Any air bubbles observed during the mixing period are removed by centrifugation. The mechanical spectra of YMG and LBG blends at three different ratios (9:1, 1:1 and 1:9, respectively) are significantly different (Figure 1.22). The blend of 9:1 ratio (90% LBG and 10% YMM, Figure 1.20) exhibited the behavior of a viscoelastic fluid. In contrast, the 1:1 and 1:9 blends exhibited gel-like properties, particularly the 1:9 blend (i.e., 10% LBG and 90% YMM, Figure 1.22) where

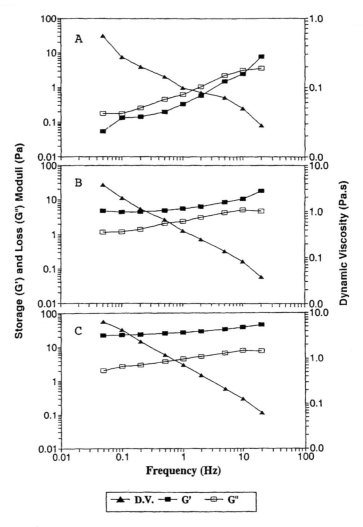

**Figure 1.22** Mechanical spectra of locust bean gum and yellow mustard mucilage blends (A: 9:1; B: 1:1 and C: 1:9) at 25°C, 0.5% (w/w) polymer concentrations. Reprinted from Reference [24] with permission from Elsevier Science.

$G'$ is much greater than $G''$, and the two moduli are almost independent of frequency. These findings suggested that, at a total polymer concentration of 0.5% (w/w), significant synergistic interactions occurred in blends of 1:1 and 1:9. The optimum ratio of synergistic interaction between YMG and LBG is 1:9, which is significantly different from the ratio of 8:2 reported by Weber and coworkers [3], possibly due to differences in sample origin and purity. Weber and coworker's YMG contained $\sim$33% fiber, 14% proteins $\sim$50% unknown material. In comparison, the material used in the Cui and coworkers' study contained 91% carbohydrates and only 4.4% protein [9]. Possible differences in the composition and molecular weight of LBG might also be contributed to the observed differences of the optimum ratios in the two studies.

In order to quantitatively compare the rheological responses of the two polymers separately as well as in blends, data from the respective rheological parameters at a fixed frequency (1 Hz) are plotted and shown in Figure 1.23. For the polymer blends, if only an additive effect occurred, the viscosity and $G'$ of the mixtures should fall between the values of LBG and YMG. The dynamic viscosity of the 9:1 blend (90% of LBG and 10% of YMM) was lower than pure LBG but slightly higher than that of YMG. In contrast, the dynamic viscosities of the other two blends (1:1 and 1:9) significantly increased (Figure 1.23). It is particularly worth noting that the mixture of 10% LBG and 90% YMM had the highest value of dynamic viscosity. Changes in storage modulus $G'$ of different blends are shown in Figure 1.23(b). The $G'$ of the 9:1 blend is 0.3 Pa and that is between the $G'$ values of LBG and YMG (0.2 and 1.7 Pa, respectively) at the same concentration. The $G'$ of blend 1:1 was three times (5.6 Pa) higher than that of YMM, while the $G'$ of blend 1:9 (10% LBG and 90% YMM) was 15 times (27 Pa) higher. This trend is in agreement with the changes in dynamic viscosity among the blends presented in Figure 1.23, thus confirming that the strongest interaction occurred in the 1:9 blend (LBG:YMM).

The phase angle ($\delta$) is another parameter used to evaluate the viscoelastic properties of polymer dispersions. An ideal liquid system shall have a phase angle of 90°, while an ideal solid system shall have a phase angle of 0°. Phase angles of real systems are in between these two extreme values. A phase angle larger than 45° would suggest that the system has a more liquid-like character. In contrast, a phase angle smaller than 45° would indicate a system with a more solid-like character. The LBG is a typical viscoelastic fluid because it has a phase angle value over 70° at 1 Hz. The phase angle value (23°) of YMG under the same experimental conditions is much smaller than 45°, suggesting that the system is more solid-like in character. In the blends of the two polymers, the phase angle of blend 9:1 is comparable to that of LBG. The phase angle of blend 1:1 is similar to that of YMG indicating that this blend is a weak gel system. The phase angle value for the 1:9 blend is smaller than that of YMG; this reveals that the 1:9 blend has a stronger gel structure and more solid-like character than that of YMG.

Interactions of LBG and YMG at 2.0% polymer concentration are examined in 0.1 M NaCl solution, as shown in Figure 1.24. The dynamic viscosities of all blends at this concentration level are higher than those of either of the polymers. The extent of the synergistic interactions increased as the ratio of LBG decreased (Figure 1.24). Similar trends are observed for $G'$, but the degree of increase is much higher for blends 1:1 and 1:9 compared to 9:1 (Figure 1.24). These

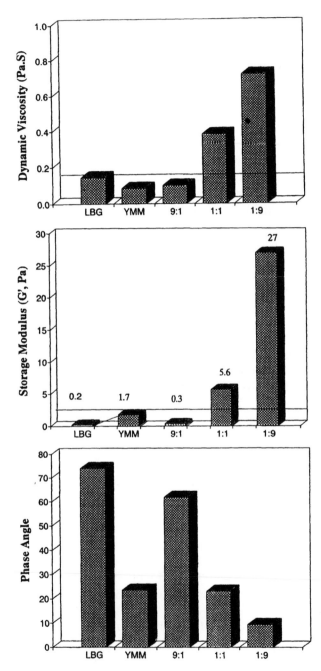

**Figure 1.23** Comparison of rheological parameters (1 Hz) of locust bean gum (LBG) and yellow mustard gum (YMG) blends and polymers in isolation (25°C, 0.5 w/w polymer concentration in water). Reprinted from Reference [24] with permission from Elsevier Science.

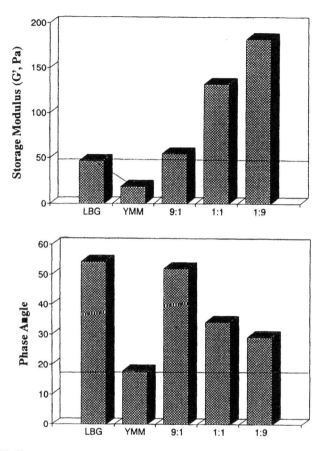

**Figure 1.24** Comparison of rheological parameters (1 Hz) of locust bean gum (LBG) and yellow mustard gum (YMG) blends and polymers in isolation (25°C, 2.0% w/w polymer concentration in 0.1 M NaCl solution). Reprinted from Reference [24] with permission from Elsevier Science.

observations are consistent with the results from 0.5% (w/w) concentration, in which the highest synergistic interactions between LBG and YMG occurred at a ratio of 1:9.

The effects of YMG/galactomannan ratio and total polymer concentration on $G'$ are demonstrated in Figure 1.25. The optimum synergism is observed at the ratio of 7:3 for YMG/LBG while highest $G'$ is obtained at the ratio of 9:1 for the YMG/fenugreek system. In the case of YMG/guar systems, synergisms are observed in a wide range of polysaccharide ratios. The extent of the synergistic interaction of YMG with the galactomannans follows the following order: YMG-LBG $\gg$ YMG-Guar > YMG-Fenugreek. Figure 1.26 shows the mechanical spectrum of the YMG-LBG system at the 0.1% level; which suggests that YMG-LBG formed a gel even at 0.1% (w/w) total polymer concentration.

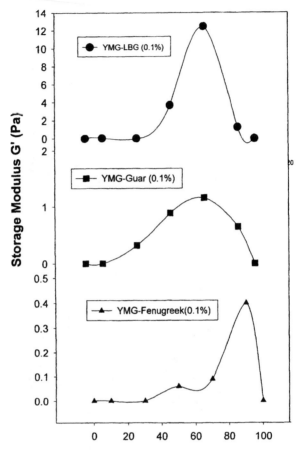

**Figure 1.25** Changes of gel strength of YMG and three galactomannans at different ratios (23.0°C, 0.113 Hz).

## 5.2. SYNERGISTIC MECHANISMS

There are a number of models proposed to interpret the synergistic phenomena between different polysaccharides. Most of the synergistic interactions between polysaccharides have been attributed to a cooperative polymer association to form "junction zones" [25]. In contrast to the cooperative binding mechanism, a mutual exclusion mechanism was recently proposed for some mixed systems in which the observed synergism, measured by an increase in viscosity or gel formation, was explained by a complex mutual exclusion effect between the two polymers as a result of polymer incompatibility. The mutual exclusion of each component from the "polymer domain" of the other is thermodynamically favorable and will increase the effective concentration of both polymers in their own microdomains [25]. Three generalized consequences of

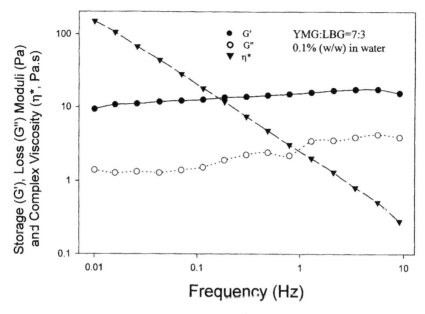

**Figure 1.26** Mechanical spectrum of a blend of yellow mustard gum and locust bean gum (7:3, 0.1% w/w) at 25°C.

mixing two different types of polymer solutions have been summarized:

(1) *Incompatibility* (mutual exclusion) results in the formation of two liquid polymer layers with each layer enriched in one of the polymers
(2) *Compatibility* results in complete miscibility and formation of a single homogeneous phase
(3) *Polymer association* (cooperative binding) results in co-precipitation of the polymers, in some instances, to the formation of a gel

The observed improvement in gelling behavior of YMG by adding small amounts of LBG at a low polymer concentration (0.5%, w/w) suggests that the synergistic interactions between LBG and YMG occurred through a cooperative association of long stretches of the two polymers into mixed "junction zones" [25]. Heating was necessary to achieve maximum interaction, which suggests that the associative synergistic interactions between LBG and YMG at low polymer concentrations required melting the ordered structures in YMG. The proposed "mixed junction" model for YMG and LBG at 0.5% (w/w) is analogous to that of mixtures of xanthan and LBG [25]. The phase angles of the three blends at 2.0% (w/w) polymer concentration were in between those of LBG and YMG (Figure 1.24) indicating that interactions of the two polymers did not improve the gelling properties of YMG at this concentration. The observed rheological

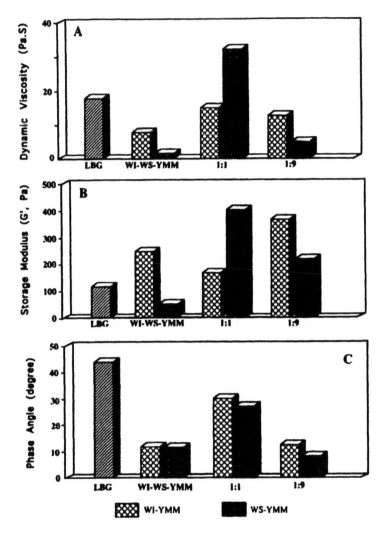

**Figure 1.27** Comparison of rheological parameters (1 Hz) of locust bean gum (LBG) and yellow mustard gum (water-insoluble, WI-YMM and water-soluble, WS-YMM, respectively) blends and polymers in isolation (25°C, 2.0% w/w polymer concentration in water). Reprinted from Reference [24] with permission from Elsevier Science.

responses at a higher polymer concentration (2.0%) could be due to the mutual exclusion or incompatibility of the two polymers. Assuming this mechanism, the increase in dynamic viscosity and storage modulus could be explained by the increase in effective concentration of both polymers due to mutual exclusion of each polymer component from the microdomains of the other [25].

In order to identify the active component of YMG that synergistically interacts with galactomannans, Cui and coworkers mixed locust bean gum with two YMG fractions, a water-soluble fraction (WS-YMG) and a water-insoluble fraction (WI-YMG) [24]. The 9:1 (LBG:YMG) blend was excluded from this test because it did not show significant interaction in a previous study. It was found that the dynamic viscosities of LBG-WI-YMG blends fell in between those of the two polymers as did the corresponding phase angles (Figure 1.27). The storage modulus $G'$ of 1:9 blend revealed a synergistic response, as it was higher than that of WI-YMG and LBG at isolation [Figure 1.27(b)]. In the blends of LBG and WS-YMG, significant synergistic interactions were observed. Dynamic viscosity and $G'$ of the blend 1:1 were much higher than those of the 1:9 blend, suggesting a shift in the optimum ratio for synergistic interaction (Figure 1.27). However, the phase angle value of the 1:1 blend was in between the two pure systems while the phase angle of the 1:9 blend had a much smaller value compared to that of WS-YMG [Figure 1.27(c)]. This finding suggested that the 1:9 blend exhibited the strongest gel structure. The mechanical spectrum of the LBG:WS-YMG system at a ratio of 1:9 is presented in Figure 1.28.

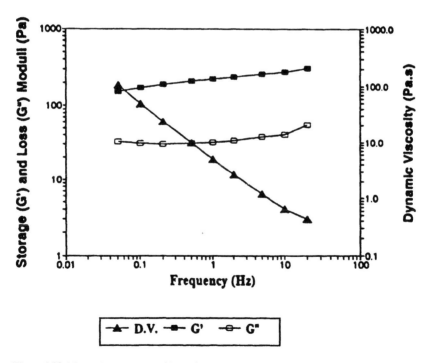

**Figure 1.28** Mechanical spectrum of locust bean gum and water-soluble yellow mustard mucilage 1:9 ratio at 25°C, 2.0% (w/w) in water. Reprinted from Reference [24] with permission from Elsevier Science.

This reflects a typical gel structure where $G'$ is much greater than $G''$ over the entire frequency range examined with the two moduli almost independent of frequency. This study confirms that the WS-YMG fraction is the major component responsible for the synergistic interactions between YMG and LBG.

In section 3.3, we described a 1,4-linked β-D-glucan backbone chain for the WSCS-I fraction. In fact, this type of structure predominated other fractions such as WSCP-I, WSCP-V and WSCS-V. In another words, the 1,4-linked β-D-glucan backbone chain is the major structure of YMG. Coincidentally, xanthan gum, a well-known microbial-produced gum that interacts with galactomannans synergistically, has a similar $(1\rightarrow 4)$-linked β-D-glucan backbone chain. It is clear that the active component of YMG is the water-soluble $(1\rightarrow 4)$-linked β-D-glucans that synergistically interact with galactomannans.

## 6. APPLICATIONS OF YMG

### 6.1. YMG AS DIETARY FIBER

Dietary fiber is defined as the endogenous components of plant material in the diet which are resistant to digestion by enzymes produced by man. They are predominantly nonstarch polysaccharides and lignins. Apparently, yellow mustard mucilage/gum falls into the dietary fiber category. Therefore, it is reasonable to expect that yellow mustard gum may also exert some physiological effects that are typical for dietary fibers, including regularizing colonic function, normalizing serum lipid levels, attenuating the postprandial glucose response and perhaps suppressing appetite. A study by Begin and coworkers examined the effects of YMG and other soluble fibers on glycemia, insulinemia and gastrointestinal function in the rat. They found that YMG, guar gum, oat β-glucan and carboxymethylcellulose significantly decreased postprandial insulin levels at 45 min, indicating a slowdown in glucose absorption. YMG decreased insulinemia primarily by delaying gastric emptying, while the other fibers increased intestinal contents and consequently decreased absorption [26]. Viscosity was considered a major contributory factor to the improved insulin status at peak time. Since viscosity of YMG increases at acid pH (see section 4), it exerts a stronger gastric effect compared to the other fibers examined. Jenkins and coworkers demonstrated that incorporating mustard fiber into white bread at levels not affecting palatability had a modest but significant effect on reducing the glycemic index of the bread in normal and diabetic human volunteers [27]. A significant reduction in percent peak rise in postprandial blood glucose was also observed (Figure 1.29). Further research on the effects of YMG in human health in addition to its applications in various foods is needed.

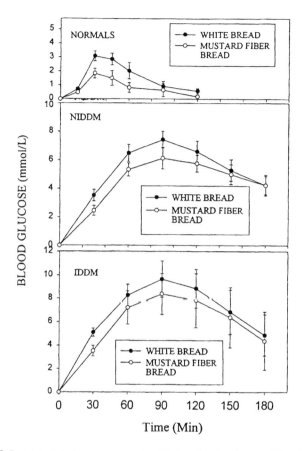

**Figure 1.29** Post-prandial glucose response to white bread (●) and mustard fiber bread (○) in normal volunteers (NORMALS), non-insulin dependent (NIDDM) and insulin-dependent (IDDM) diabetics (Adapted from Reference [27]).

## 6.2. YMG AS STABILIZER

The applications of gums/hydrocolloids in food systems are based on their functional properties. In sections 4 and 5, YMG demonstrated similar properties to those of xanthan gum, one of the most widely used gums in the food industry. Similar applications can be expected for YMG, including synergistic interactions with galactomannans. For example, salad cream products using a YMG:LBG (9:1) blend (at 0.3% w/w gum concentration) exhibited favourable stability and rheological properties compared to commercial products containing xanthan gum with or without alginates (Figure 1.30, Table 1.19) [28].

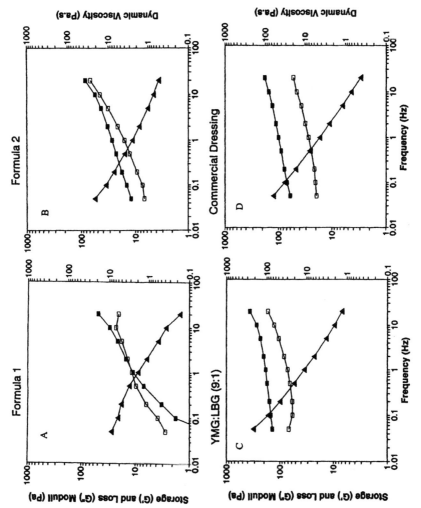

**Figure 1.30** Comparison of mechanical spectra of salad creams: A, no YMG added; B, 0.3% YMG; C, 0.3% YMG:LBG:blend (9:1) and D, commercial salad dressing. Reprinted from Reference [90] by permission of Oxford

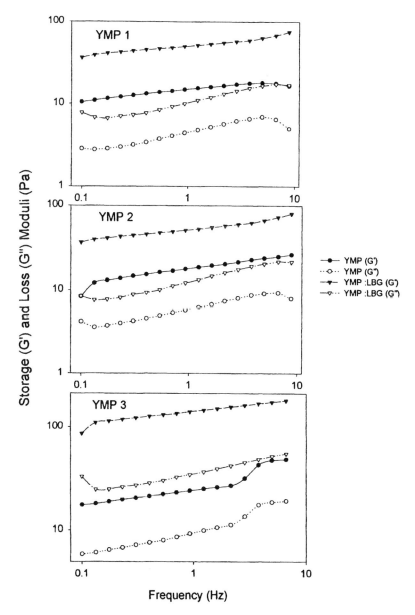

**Figure 1.31** Mechanical spectra of yellow mustard powder alone and after addition of LBG. YMP1-3 are three commercial yellow mustard powders.

## Synergistic Effects of Ymp:Lbg on Processed Chicken Meat

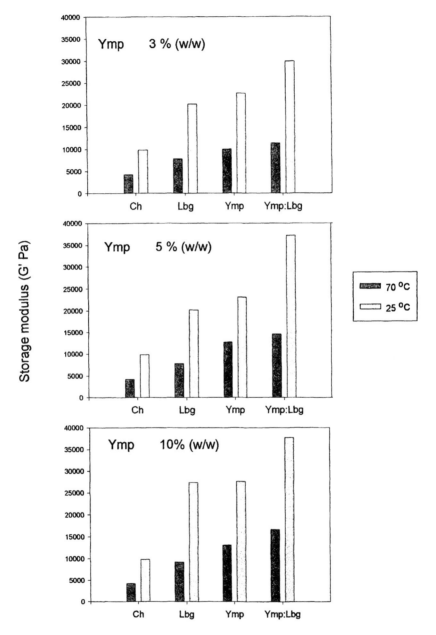

**Figure 1.32** Effect of addition of three levels of yellow mustard powder with locust bean gum on the gel strength of chicken meat.

**TABLE 1.19. Emulsion Stability of Salad Creams.**

| Salad Cream | Stabilizer (Level) | Emulsion Stability (%) |
|---|---|---|
| Salad Cream Based on Table 1: | | |
| Cream 1 | No gum added | 58.6 ± 0.1[a] |
| Cream 2 | YMG (0.3%) | 72.3 ± 0.4 |
| Cream 3 | YMG:LBG blend (9:1)(0.3%) | 65.5 ± 0.4 |
| Commercial Salad Creams[b] : | | |
| A | Xanthan gum and alginate | 67.8 ± 1.5 |
| B | Xanthan gum and alginate | 68.7 ± 0.2 |
| C | Xanthan gum | 80.1 ± 0.1 |

[a]$n = 2$, mean ± S.D.
[b]The level of stabilizers in the commercial products was unknown.
Reprinted from Reference [28] by permission of Oxford University Press, © 1996.

A commercial salad dressing produced in New Zealand was stabilized with yellow mustard mucilage.

There are substantial amounts of commercial ground yellow mustard products that contain ~5% YMG and are used in processed meats as condiments or bulking agents. Application of the synergism described in section 5 could improve the rheological characteristics of the final product and possibly the shelf life. For example, the addition of a small amount of LBG to commercial yellow mustard flour significantly increased the gelling strength (Figure 1.31). The application of LBG added to yellow mustard flour to meat systems enhanced the rheological characteristics of the final products (Figure 1.32).

Currently, yellow mustard gum is commercially produced in Canada for food and cosmetic applications [11]. A skin-care moisturizing lotion was prepared using YMG according to a formula including 15% YMG preparation [11]. The prepared moisturizing lotion has a yield stress of 11.6 Pa and shear thinning flow behavior with a favorable handfeel compared to lotions prepared with other commercial gums [11]. Additional applications of YMG in the food and cosmetic industries are possible, however, further research and development are needed.

# 7. REFERENCES

1. Hemmingway J. S. Mustard and condiment products. *Encyclopaedia of Food Science and Technology and Nutrition.* New York: Academic Press; 1993. pp. 3178–82.
2. Cui W., Eskin M. N. A. Processing and properties of mustard products and components. Mazza G., Editor. *Functional Foods: Biochemical and Processing Aspects.* Lancaster, Basel: Technomic Publishing Co., Inc.; 1998. pp. 235–64.

3. Weber F. E., Taillie S. A., Stauffer K. R. Functional characteristics of mustard mucilage. *J. Food Sci.* 1974; 39:461–6.

4. Bailey K., Norris F. W. The nature and composition of the mucilage from the seed of white mustard (*Brassica alba*). *Biochem. J.* 1932; 26:1609–23.

5. Hirst E. L., Rees D. A., Richardson N. Seed polysaccharides and their role in germination. *Biochem. J.* 1965; 95:453–8.

6. Woods D. L., Downey R. K., Mucilage from yellow mustard. *Can. J. Plant Sci.* 1980; 60:1031–3.

7. Siddiqui I. R., Jones J. D., Kalab M., Yiu S. H. Mucilage in yellow mustard (*Brassica hirta*) seeds. *Food Microstructure* 1986; 5:157–62.

8. Vaughan J. G. *The Structure and Utilization of Oilseeds.* London: Chapman and Hall Ltd.; 1970.

9. Cui W., Eskin N. A. M., Biliaderis C. G. Chemical and physical properties of yellow mustard (*Sinapis alba* L) mucilage. *Food Chem.* 1993; 46:169–76.

10. Vose J. R. Chemical and physical studies of mustard and rapeseed coats. *Cereal Chem.* 1974; 51:658–65.

11. Cui W., Eskin M., Han N., Duan Z., Zhang X. Extraction process and use of yellow mustard gum. Patent pending, Canada 2270750, 1999.

12. Theander O., Aman P., Miksche G. E., Yasuda S. Carbohydrates, polyphenols and lignin in seed hulls of different colors from turnip rapeseed. *J. Agric. Food Chem.* 1977; 25:270–3.

13. Cui W., Eskin N. A. M. Yellow mustard gum: Optimization of extraction process and rheological properties. *Food Hydrocolloids* (Submitted) 2000.

14. Cui W., Eskin N. A. M., Biliaderis C. G. Water-soluble yellow mustard (*Sinapis alba* L) polysaccharides—partial characterization, molecular size distribution and rheological properties. *Carbohydrate Polymers* 1993; 20:215–25.

15. Cui W. W., Eskin N. A. M., Biliaderis C. G. Fractionation, structural-analysis, and rheological properties of water-soluble yellow mustard (*Sinapis alba* L) Polysaccharides. *Journal of Agricultural and Food Chemistry* 1994; 42:657–64.

16. Cui W. W., Eskin M. N. A., Biliaderis C. G., Marat K. NMR characterization of a 4-*O*-methyl-beta-D-glucuronic acid-containing rhamnogalacturonan from yellow mustard (*Sinapis alba* L) mucilage. Carbohydrate Research 1996; 292:173–83.

17. Cui W., Eskin N. A. M., Biliaderis C. G. NMR characterization of a water-soluble 1,4-linked beta-D-glucan having ether groups from yellow mustard (*Sinapis alba* L) mucilage. Carbohydrate Polymers 1995; 27:117–22.

18. Aspinall G. O. *Polysaccharides.* Toronto, Canada: Pergamon Press; 1970.

19. Wallingford L., Labuza T. P. Evaluation of the water binding properties of food hydrocolloids by physicochemical methods and in a low fat meat emulsion. *J. Food Sci.* 1983; 48:1–5.

20. Yasumatsu K., Sawada K., Moritaka S., Misaki M., Toda J., Wada T., Ishii K. Whipping and emulsifying properties of soybean products. *Agric. Biol. Chem.* 1972; 36:719–27.

21. Witcomb P. J., Gutowski J., Howland W. W. Rheology of guar solutions. *J. Applied Polymer Sci.* 1980; 25:2815–27.

22. Navarini L., Cesaro A., Ross-Murphy S. B. Exopolysaccharides from *Rhizobium muliloti* YE-2 grown and different osmolarity conditions: viscoelastic properties. *Carbohydr. Res.* 1992; 223:227–34.

23. Anguilar C. A., Ziegler G. R. Effect of temperature and electrolytes on the viscosity of aqueous dispersions of mustard seed (*Sinapsis alba*) mucilage. *Food Hydrocolloids* 1990; 4:161–6.

24. Cui W., Eskin N. A. M., Biliaderis C. G., Mazza G. Synergistic interactions between yellow mustard polysaccharides and galactomannans. Carbohydrate Polymers 1995; 27:123–7.

25. Morris E. R. *Mixed Polymer Gels: Food Gels*. Harris P., ed. London and New York: Elsevier Applied; 1990.

26. Begin F., Vachon C., Jones J., Wood P., Savoie L. Effect of dietary fibers on glycemia and insulinemia and on gastrointestinal function in rats. *Can. J. Physiol. Pharmacol.* 1988; 67:1265–71.

27. Jenkins A. L., Jenkins D. J. A., Ferrari F., Collier G., Rao A. V., Samuels S., Jones J. D., Wong G. S., Josse R. G. Effect of mustard seed fiber on carbohydrate tolerance. *J. Clin. Nutr. Gastroenterol.* 1987; 2:81–6.

28. Cui W., Eskin N. A. M. Interaction between yellow mustard gum and locust bean gum: impact on a salad cream product. *Gums and Stabilisers for the Food Industry 8*. Phillips G. O., Willians P. A., Wedlock D. J., eds. IRL Press at Oxford University; 1996. pp. 161–70.

# CHAPTER 2

# Flaxseed Gum

## 1. INTRODUCTION

**F**LAX has been used for human consumption for over five thousand years. The common flax plant (*Linum usitatissimum*) is a member of a small family, the Linaceae, which includes about a dozen genera and some 150 species that are widely distributed in the temperate and subtropical regions of the world [1]. The seed of the flax plant is known as linseed or flaxseed. Historically, the flax crop has produced oil from the seed (known as linseed oil) and fiber from the stem. The ex-Soviet Union was the most important producer of fiber-type flax (65–70% of world production), whereas Canada is the largest producer and exporter of oil-type flaxseed. The Canadian share of the world flaxseed production is about 31–35% [1]. The production of flaxseed has declined in the last twenty years in North America due to decreased demand for linseed oil in the painting industry [2]. Extensive research has been carried out during the last 10 years to explore other uses of flaxseed, particularly food uses [2]. Flaxseed mainly contains oil (42%), protein (30%), soluble and insoluble dietary fiber (9% and 20%, respectively) and anticarcinogenic lignans (0.3%) [3,4]. The properties and applications of flaxseed oil, protein and lignans are described elsewhere [2,3,5–7]. The focus of the present chapter is on the soluble fiber portion of flaxseed: the most recent advances in

extraction processes, chemical structure, functional properties and applications
of flaxseed gum.

## 2. EXTRACTION AND PROCESSING OF FLAXSEED GUM

### 2.1. OCCURRENCE

Flaxseed has a flat and oval shape pointed on one end and is on average 5 mm
in length, 2.5 mm in width and 1.5 mm in thickness [8]. The surface of the seed
is smooth, highly polished and varies in color from yellow to dark brown. The
seed coat (testa) consists of five distinct layers: the mucilage (epidermis) is in
the outermost layer, followed by round cells, fibers, cross cells and pigment
cells, as demonstrated in Figure 2.1 [8]. A scanning electron microscopic graph
of the testa layers and part of the endosperm layer of a dry flaxseed (Norman)
is shown in Figure 2.2. A thick mucilage layer and a pigment layer are clearly
identified, while the round cells, fibers and the cross cell are less distinguishable.
The mucilage is secondary wall material and is striated in the cell in dry seeds.
However, the mucilage expands rapidly upon hydration. It lifts the cuticle and
outer epidermal wall and essentially leads to the breakage of the mucilage cells

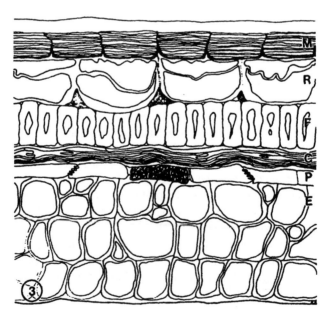

**Figure 2.1** Composite line drawing showing the five layers of flaxseed coat (900×). Abbreviations:
M, mucilage; R, round cell; C, cross cell; F, fiber; P, pigment cell; and E, endosperm. Reprinted
from Reference [8] with permission from AOCS Press.

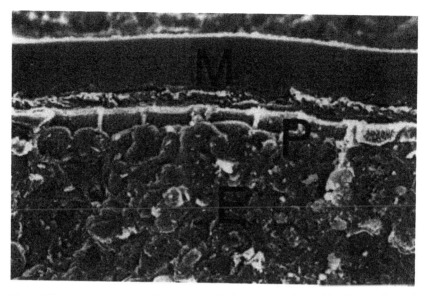

**Figure 2.2** A scanning electron microscopy graph of test layers and a partial endosperm layer of a dry flaxseed.

under the pressure of expansion [8]. When there is a sufficient amount of water, the mucilage will be solubilized and, therefore, extracted into water, as will be described in detail in the following section.

## 2.2. EXTRACTION OF GUM FROM FLAXSEED

A characteristic of flaxseed gum is that it is easy to dissolve in cold water. This characteristic allows the use of mild extraction conditions to extract the gum from raw materials. A typical extraction procedure is to soak the seeds in water at various temperatures with or without stirring for a period of 3 to 16 hr. The extract is filtered by nylon cloth, cheesecloth or other means to be seed free and is then added to 2–3 times its volume of organic solvents (ethanol or acetone) to precipitate the gum. The precipitated gum is washed with the same solvent, redissolved in water, sometimes dialyzed against distilled water and then dried.

Erskine and Jones stirred 500 g of flax seed in water (1 L) for 1–2 min, and the liquid was immediately decanted through a Buchner funnel (without filter paper) [9]. The seeds were washed on the filter with 1 L water and then stirred in 5 L water for 3 hr. The mucilage solution was freed from seeds and debris by filtration. Muralikrishna and co-workers extracted flaxseed gum by soaking the seeds in water (1:6, w/v) for 24 hr with occasional stirring [10]. After decanting the mucilage was precipitated with alcohol and dried to give a

TABLE 2.1. **Influence of Time and Temperature on Extraction of Flaxseed Gum from Linott Seed.**

| Extraction Time (hr) | Yield (% Seed) | | |
|:---:|:---:|:---:|:---:|
| | 25°C | 100–25°C[a] | 100°C |
| 0.5 | 3.6 ± 0.1[b] | 6.5 ± 0.1 | 8.4 ± 0.2 |
| 1.0 | 4.4 ± 0.3 | 7.2 ± 0.3 | 8.5 ± 0.2 |
| 2.0 | 4.8 ± 0.3 | 7.6 ± 0.2 | 8.6 ± 0.2 |
| 4.0 | 5.2 ± 0.4 | 7.8 ± 0.3 | 8.9 ± 0.6 |
| 6.0 | 5.2 ± 0.3 | 7.9 ± 0.4 | 8.9 ± 1.0 |
| 8.0 | 5.3 ± 0.3 | 8.2 ± 0.5 | 8.9 ± 1.2 |

[a]Extraction performed at 25°C with water initially at 100°C.
[b]Standard deviation, $n = 4$.
Reprinted from Reference [13] with permission from IFT.

yield of 6%. Similar methods were also reported by Susheelamma [11] and Wannerberger and coworkers [12]. Mazza and Biliaderis examined the effects of temperature and extraction time on the yield and composition of flaxseed gum [13]. It was found that higher temperature increased the gum yield from 5% to 9% (Table 2.1), but the extracted gum was contaminated with high levels of proteins [13]. Fedeniuk and Biliaderis later confirmed that higher temperature resulted in higher yield but with 20% to 22% protein contaminants (Table 2.2) . After treatment with Vega clay, the yield and protein content

TABLE 2.2. **Effect of Extraction and Purification Procedures on Yield and Purity of Linseed Mucilage.**

| Extraction[a] Condition | Yield[b] (%) | Protein[c] (%) | Ash[c] (%) |
|:---|:---:|:---:|:---:|
| Linott 25°C, 2 h, EtOH ppt | 4.0 ± 0.4[d] | ND[e] | ND |
| Linott 100–25°C, 2 h, EtOH ppt | 5.1 ± 0.4 | ND | ND |
| Linott 80°C, 2 h, EtOH ppt | 8.4 ± 0.4 | 22.8 ± 0.6 | 6.3 ± 0.1 |
| Linott 4°C, 24 h, lyophilized[e] | 3.6 ± 0.1 | <1 | 7.7 ± 0.1 |
| Linott 80°C, 2 h, lyophilized[e] | 6.0 ± 0.1 | 19.7 ± 0.6 | 4.4 ± 0.1 |
| Vega clay treated | 4.7 ± 0.3 | 4.3 ± 0.1 | 8.7 ± 0.1 |
| meal 55°C, 0.5 h, EtOH ppt | 9.1 ± 0.2 | 29.2 ± 0.2 | 5.7 ± 0.1 |
| meal 80°C, 0.5 h, EtOH ppt | 9.4 ± 0.1 | 28.6 ± 0.4 | 4.2 ± 0.1 |
| meal 55°C, 0.5 h, lyophilized[e] | 9.5 ± 0.4 | 28.3 ± 0.2 | 3.4 ± 0.3 |

[a]Source, temperature and time of extraction, and method of isolation.
[b]% basis of seed or meal.
[c]% on a dry mucilage basis.
[d]Means ± SD ($n = 3$).
[e]Not determined.
Reprinted from Reference [14] with permission from the *Journal of Agricultural and Food Chemistry.*
© 1994 The American Chemical Society.

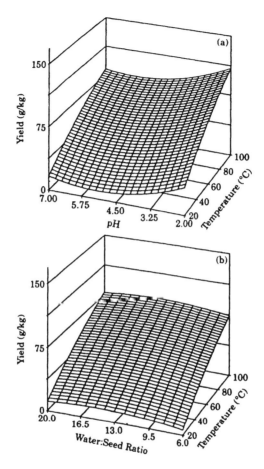

**Figure 2.3** Response surfaces for the effect of temperature and pH (a, water:seed ratio 13), temperature and water:seed ratio (b, pH 6.5) on yield of flaxseed gum. Reprinted from Reference [16] with permission from *Food Science and Technology—Lebensmittel-Wissenschaft & Technologie.* © 1994 Academic Press Ltd.

were comparable with gums extracted at room temperature (4.7% yield and 4.3% proteins) [13,14].

Flaxseed gum can also be extracted by treating flaxseed with superheated steam. Sanftleben treated flaxseed with wet steam under pressure [15]. The liquid extract was passed into a receptacle under a substantially lower pressure [15].

Since earlier reported extraction methods are either time consuming or resulted in high levels of protein contaminants [13–15], Cui and co-workers optimized the water extraction process using Response Surface Methodology to obtain gums of maximum yield and minimum impurities [16]. Extraction

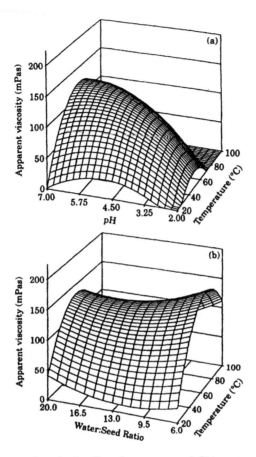

**Figure 2.4** Response surfaces for the effect of temperature and pH (a, water:seed ratio 13), temperature and water:seed ratio (b, pH 6.5) on the apparent viscosity of flaxseed gum (apparent viscosity was determined at 25°C, pH 6.5, polymer concentration 1.0% and shear rate 46.16/s). Reprinted from Reference [16] with permission from *Food Science and Technology—Lebensmittel-Wissenschaft & Technologie*. © 1994 Academic Press Ltd.

temperature (25–100°C), pH (2.0–7.0), water:seed ratio (6.0–20.0) and extraction time were the factors investigated with respect to yield, apparent viscosity and protein content. The effects of temperature, pH and seed:water ratio on the yield of flaxseed gum are shown in Figure 2.3. Temperature is the major factor identified affecting the yield of flaxseed gum. As temperature increases, the yield of gum increases almost linearly, irrespective of the pH and/or water:seed ratio. However, the highest viscosity is obtained from pH 5.0–7.0 and temperatures between 50–80°C when the water:seed ratio is constant at 13 (Figure 2.4). The effects of independent variables on the protein content of the gums are illustrated in Figure 2.5. At a constant water:seed ratio of 13.0, high protein

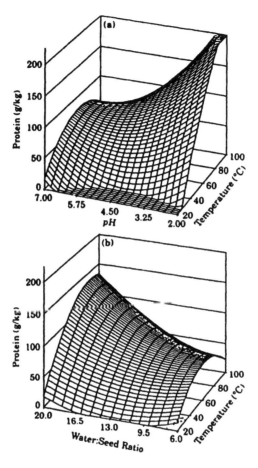

**Figure 2.5** Response surfaces for the effect of temperature and pH (a. water:seed ratio 13), temperature and water:seed ratio (b, pH 6.5) on the protein content of flaxseed gum. Reprinted from Reference [16] with permission from *Food Science and Technology—Lebensmittel-Wissenschaft & Technologie.* © 1994 Academic Press Ltd.

content gums are obtained at high temperatures ($\sim$100°C) and low pHs ($\sim$2.0); under these conditions, the protein contents of gum could reach 20% or higher. In summary, extraction conditions have a significant effect on the yield, protein content and rheological properties of flaxseed gum. Of the four parameters examined, temperature and pH are found to be the major factors affecting yield and quality (low amounts of protein contaminants and high viscosity). Optimum extraction conditions are obtained at a temperature of 85–90°C, a pH of 6.5–7.0, a water:seed ratio of 13 and an extraction time of 2.5–3 hr. This optimum processing condition gives a relatively high yield of gum (ca. 8%) with low levels of protein contaminants (<8%).

## 2.3. EXTRACTION OF GUM FROM FLAXSEED MEAL

Because linseed oil is the primary product of flaxseed, a large amount of flaxseed meal is produced as a by-product from the oil crashing industry. Therefore, it is a viable approach to extract gum from the meal. Solvent-extracted meal could be separated by air and/or screened to obtain kernel and hull fractions [17]. The hull fraction is then extracted with water (water/solid ratio of 30:1, pH of 4.5 and temperature of 60°–80°C) [17]. The extracted liquid is centrifuged, adjusted to a pH of 7, concentrated by evaporation and spray dried. Mason and Hall also described a process for preparating flaxseed gum by extracting the meal cake with 5% sodium chloride, followed by centrifugation with activated carbon, vacuum concentration and alcohol precipitation [17]. The product was described as a useful emulsifying agent for chocolate milk [17]. Seed or meal cake was also immersed in cold or warm water containing an iron salt to prevent extraction of tannin pigment [18]. Gums extracted from flaxseed meals frequently contain high levels of proteins. For example, at 55°C and 80°C for 30 min, water extraction of flaxseed meal gave a yield of 9–9.5% with about 29% protein contaminants (Table 2.2) [14].

## 2.4. EXTRACTION OF GUM FROM FLAXSEED HULL

Flaxseed could be separated into a hull fraction ($\sim$37%) and a kernel fraction (63%) [19,20]. Because flaxseed gum (mucilage) is deposited in the outer layer of the seed coat (Figures 2.1 and 2.2), it would be economically viable to extract the gum from the hull fraction and utilize the kernel fraction for other purposes. A novel dehulling process of flaxseed developed by Cui and Mazza made this approach possible [19]. This dehulling process is easy to scale up to produce flaxseed hulls and kernels in a large quantity [19]. The flaxseed kernel has an attractive golden color and could be used as a novel ingredient for a number of food products, such as breakfast cereals, bakery products and snack foods. The hull fraction, which is rich in fiber and lignans, is first extracted with organic solvent to extract the lignans and some phenolic components and then, with water is extracted under the optimum conditions described earlier [16,19]. A yield of 18% of flaxseed gum is produced from the hull fraction with acceptable purity.

## 3. STRUCTURE AND CHEMICAL PROPERTIES OF FLAXSEED GUM

### 3.1. CHEMICAL AND MONOSACCHARIDE COMPOSITIONS

Flaxseed gum contains 50–80% carbohydrates, 4–20% proteins and 3–9% ash. These compositions vary with varieties/raw materials and extraction

TABLE 2.3. **Comparison of Analytical Data of Flaxseed Gum and Commercial Gums.**

| | Flaxseed Gums | | | | Commercial Gums | | |
|---|---|---|---|---|---|---|---|
| | **Norman** | **Omega** | **Foster** | **84495** | **Arabic** | **Guar** | **Xanthan** |
| Loss on drying (105°C; %)[a] | 6.5 | 3.7 | 14.4 | 11.5 | 12.8 | 8.6 | 10.2 |
| Total ash (55°C; %) | 7.4 | 8.2 | 8.4 | 3.3 | 1.2 | 11.9 | 1.5 |
| Nitrogen (Kjeldahl, %) | 1.5 | 2.69 | 2.95 | 2.42 | 0.34 | 1.31 | 0.86 |
| Uronic acid (%) | 21.0[b] | 25.1 | 23.9 | 15.7 | 15.0[c] | 0.0 | 21.5[d] |
| Relative neutral sugar composition (%) | | | | | | | |
| Rhamnose | 21.2 | 27.2 | 25.6 | 12.8 | 34.0 | 0.0 | 0.0 |
| Fucose | 5.0 | 7.1 | 5.8 | 3.0 | 0.0 | 0.0 | 0.0 |
| Arabinose | 13.5 | 9.2 | 11.0 | 18.1 | 24.0 | 0.0 | 0.0 |
| Xylose | 37.4 | 28.2 | 21.1 | 42.5 | 0.0 | 0.0 | 0.0 |
| Galactose | 20.0 | 24.4 | 28.4 | 18.4 | 45.0 | 33.0 | 0.0 |
| Glucose | 2.1 | 3.6 | 8.2 | 3.7 | 0.0 | 0.0 | 50.7 |
| Mannose | 0.0 | 0.0 | 0.0 | 0.0 | 0.0 | 67.0 | 49.3 |

[a]On a dry weight base.
[b]Galacturonic acid for flaxseed gum.
[c]Content of glucuronic acid.
[d]Glucuronic acid content and monosaccharide composition of xanthan gum.
Reprinted from Reference [26] with permission from Elsevier Science.

conditions [10–14,21,22]. The polysaccharide portion of flaxseed gum is comprised of seven monosaccharides and uronic acids. A comparison of moisture, ash, protein contents and monosaccharide composition of flaxseed gum is presented in Table 2.3. Large variations of monosaccharide composition of different varieties/cultivars were also reported by other researchers [12,20,21,23]. For example, Wannerberger and coworkers reported that the content of uronic acid varied from 21% to 36% in gums from 23 flax varieties grown in Sweden [12]. A comparison of monosaccharide composition and uronic acid content of six of the 23 varieties studied is presented in Table 2.4. The monosaccharide composition and galacturonic acid content of flaxseed gums extracted from six brown and six yellow flaxseed cultivars grown in Canada are summarized in Table 2.5. In the brown seed group, gum extracted from Royal contained the lowest amounts of rhamnose (10.7%) and galacturonic acid (13.9%), followed by Reina, Norman, 22-87-2-159, Verne and Atlante. In the yellow flaxseed group, cultivars including 84495, 92-5103, APF-9006 and APF-9007 contained a significantly lower amount of acidic fraction (12.9–14.5% rhamnose, 13.8–16.2% galacturonic acid) but a much higher amount of neutral polysaccharides (40–48% xylose) [21]. This observation corresponds well with their

**TABLE 2.4. Relative Composition of the Water-Soluble Polysaccharides in Mucilage Extracted from Liflora, Linda, Ariadna, Belinka and Szegedi 62 (%) [12].**

|  | Liflora | Linda | Ariadna | Belinka | Szegedi 62 |
|---|---|---|---|---|---|
| Nonstarch polysaccharides (g/100 g) | 50·2 | 65·5 | 62·7 | 64·3 | 53·9 |
| Total carbohydrates (g/100 g) | 69·7 | 70·4 |  |  |  |
| Relative composition (%) |  |  |  |  |  |
| Rhamnose + fucose | 16 | 15 | 15 | 11 | 11 |
| Arabinose | 8·0 | 9·0 | 13 | 10 | 12 |
| Xylose | 19 | 22 | 23 | 29 | 38 |
| Mannose | 1·0 | 1·0 | 2·0 | 1·0 | 1·0 |
| Galactose | 16 | 12 | 12 | 13 | 12 |
| Glucose | 4·0 | 6·0 | 4·0 | 4·0 | 5·0 |
| Uronic acids | 36 | 35 | 32 | 32 | 21 |

rheological properties; all gums extracted from these cultivars exhibited weak gel properties (see section 4). In contrast, gums extracted from Foster and Omega seed contained a substantially higher amount of acidic fraction (25.6–27.3% rhamnose, 23.9–25.1% galacturonic acid, respectively) and a much lower content of neutral polysaccharides (21.1–28.2% xylose). These two gums exhibited weaker rheological properties compared to others from the same color group [21].

On examination, 109 accessions of flaxseed from 12 geographical regions including oil, fiber, brown- and yellow-seeded types were analyzed for water-soluble polysaccharides and sugar composition [23]. The content of gum ranged from 3.6% to 8.0%. The ratio of rhamnose to xylose, indicative of the ratio of acidic to neutral polysaccharides, ranged from 0.3 to 2.2 with an overall mean of 0.7 (Table 2.6). The high degree of variation of the rhamnose/xylose ratio suggests that the gum from flaxseed accessions has diverse characteristics covering the spectrum from acidic to neutral polysaccharides types.

## 3.2. STRUCTURE/MOLECULAR WEIGHT HETEROGENEITY OF FLAXSEED GUM

As described in the previous section, flaxseed gum is comprised of seven monosaccharides and 13–36% uronic acids [12,21]. Two types of polysaccharides are made from those sugars and uronic acid: a neutral polysaccharide

TABLE 2.5. Relative Monosaccharide Composition and Uronic Acid Content of Gums Extracted from Six Yellow and Six Brown Flaxseed Cultivars.

| Monosaccharide and Uronic Acid (%) | Yellow Flaxseed | | | | | | Brown Flaxseed | | | | | |
|---|---|---|---|---|---|---|---|---|---|---|---|---|
| | 84495 | 92-5103 | APF-9006 | APF-9007 | Foster | Omega | Norman | Royal | 22-87-2-159 | Reina | Verne | Atlante |
| Rhamnose[a] (%) | 12.8 | 14.4 | 13 | 13.7 | 25.6 | 27.2 | 21.2 | 10.7 | 24.5 | 13.6 | 25.3 | 25.8 |
| Fucose[a] | 3 | 4.3 | 4.3 | 5.2 | 5.8 | 7.1 | 5 | 3.4 | 5.5 | 3.5 | 5.7 | 5.9 |
| Arabinose[a] | 18.1 | 16.7 | 16.7 | 16.5 | 11 | 9.2 | 13.5 | 14.8 | 13.9 | 16.5 | 8.7 | 9.9 |
| Xylose[a] | 42.5 | 39 | 48.7 | 45.2 | 21.1 | 28.2 | 37.4 | 39.7 | 24 | 44.4 | 28.3 | 23.4 |
| Galactose[a] | 18.4 | 15.1 | 13.6 | 15 | 28.4 | 24.4 | 20 | 16.4 | 27 | 19.1 | 21.4 | 25.6 |
| Glucose[a] | 3.7 | 9.9 | 4.3 | 4 | 8.2 | 3.6 | 3.1 | 13.8 | 4.3 | 2.6 | 9 | 7.9 |
| Galacturonic acid[b] | 15.7 | 15.4 | 13.8 | 16.2 | 23.9 | 25.1 | 23.2 | 13.9 | 21.1 | 15 | 24 | 21 |

[a]Relative monosaccharide composition is reported.
[b]Galacturonic acid content on gum weight basis.
Reprinted from Reference [21] with permission from Elsevier Science.

69

TABLE 2.6. **Distribution of Water-Soluble Polysaccharides (Relative Percent) in Flaxseed.**

| Component | Minimum | Maximum | Mean | Sd[a] | Cv[a] |
|---|---|---|---|---|---|
| Carbohydrate content | 3.6 | 8.0 | 6.2 | 0.8 | 12.4 |
| L-Rhamnose | 8.2 | 23.8 | 14.2 | 3.1 | 21.4 |
| L-Fucose | 1.9 | 9.1 | 3.5 | 0.9 | 25.2 |
| L-Arabinose | 3.2 | 15.4 | 10.6 | 2.0 | 18.6 |
| D-Xylose | 10.9 | 32.0 | 23.2 | 4.4 | 18.8 |
| D-Galactose | 12.7 | 26.9 | 19.1 | 2.7 | 14.0 |
| D-Glucose | 21.3 | 40.0 | 28.9 | 3.6 | 12.5 |
| Rhamnose/xylose ratio | 0.3 | 2.2 | 0.7 | 0.3 | 44.9 |
| Arabinose/xylose ratio | 0.1 | 0.9 | 0.5 | 0.1 | 22.5 |

[a]Coefficient of variation in percent.

Reprinted from Reference [23] with permission from the *Journal of Agricultural and Food Chemistry.* © 1995 The American Chemical Society.

mainly composed of xylose, arabinose and galactose and an acidic polysaccharide composed of D-galactose, L-rhamnose and D-galacturonic acid.

The neutral arabinoxylan has a $(1 \rightarrow 4)$-$\beta$-D-xylosyl backbone to which arabinose and galactose side chains are attached at positions 2 and/or 3. The acidic polysaccharide has a backbone of $(1 \rightarrow 2)$-linked $\alpha$-L-rhamnopyranosyl and $(1 \rightarrow 4)$-linked D-galactopyranosyluronic acid residues, with side chains of fucose and galactose, the former essentially at the non-reducing end. The ratio of L-rhamnose, L-fucose, L-galactose and D-galacturonic acid is 2.6:1:1.4:1.7 [9,10].

The two types of polysaccharides of flaxseed gum can be fractionated by ion exchange chromatography, precipitation of the acidic polysaccharide with some complexing agents or a combination of both [9,10,14,22,24]. Cetavlon, cetyltrimethylammonium bromide, has been used to separate acidic polysaccharides from their neutral counterparts [10,25]. Hunt and Jones separated the acidic polysaccharide from the neutral counterpart of flaxseed gum by repeated precipitation with Cetavlon [25]. The acidic polysaccharide was further fractionated into two fractions with cupric acetate solution (7%): a copper-insoluble fraction (CuI) and a copper-soluble fraction (CuII). The structures of the two acidic polysaccharides and the neutral polysaccharides were elucidated by methylation analysis and periodate oxidation.

In studies by Muralikrishna and coworkers, Cetavlon was added to 0.5% mucilage solutions to obtain precipitable (CeP, 58%) and non-precipitable (CeNP, 28%) fractions [10]. The CeP fraction was further fractionated by DEAE-cellulose ($CO_3^{2-}$ form) eluted with water, 0.2 M ammonium carbonate and 0.3 M NaOH, sequentially. The CeNP fraction was not completely soluble in water, therefore, it was separated into water-soluble (65%) and

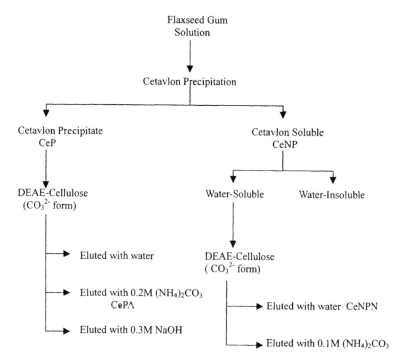

**Figure 2.6** Fractionation flowchart of flaxseed gum. Adapted from Reference [10] with permission.

water-insoluble (35%) portions. The water-soluble portion was further fractionated by DEAE-cellulose ($CO_3^{2-}$ form) eluted with water and 0.1 M ammonium carbonate [10]. The flowchart of the fractionation of flaxseed gum is demonstrated in Figure 2.6. The yield and monosaccharide composition of all fractions are presented in Table 2.7. Of those fractions, an acidic polysaccharide CePA (0.2 M ammonium carbonate eluted fraction of CeP) and a neutral polysaccharide CeNPN (water-eluted fraction of CeNP) were found to be homogeneous by cellulose acetate membrane electrophoresis, gel filtration chromatography and untracentrifugation [10]. CeNPN is composed of L-arabinose, D-xylose and D-galactose in the ratio of 3.5:6.1:1. Methylation analysis revealed that CeNPN was a highly branched arabinoxylan. The ratios of nonsubstituted to single-substituted (2-substituted) and double-substituted (2,3-substituted) D-xylosyl residues are 10.58:1:2.55. All the D-galactosyl, most L-arabinosyl and some D-xylosyl residues were non-reducing end sugars from the side chain [10], as shown in Table 2.8. Partial hydrolysis of CeNPN gave a degraded polysaccharide consisting mainly of xylose. Methylation analysis of the degraded polysaccharide suggested $(1\rightarrow4)$-linked xylosyl residues arising from a linear xylan chain (Table 2.8). Partial characterization

TABLE 2.7. Sugar Composition (%) of Flaxseed Gum Fractions Derived from Linseed Mucilage.

| Fraction | Yield (%) | Neutral Sugars | | | | | | GalA[a] |
|---|---|---|---|---|---|---|---|---|
| | | Rha | Fuc | Ara | Xyl | Gal | Glc | |
| Cetavlon—precipitable | 58 | 37 | 12 | 6 | ... | 22 | 2 | 21 |
| DEAE-cellulose eluted with 0.2 M (NH$_4$)$_2$CO$_3$ (CePA) | 62 | 37 | 14 | ... | ... | 21 | ... | 28 |
| DEAE-cellulose eluted with 0.3 M NaOH | 30 | 36 | 9 | 30 | ... | 17 | 8 | ND |
| Cetavlon—non-precipitable | 28 | ... | ... | 26 | 53 | 15 | 3 | 3 |
| Non-precipitable water-soluble | 65 | ... | ... | 28 | 52 | 16 | 2 | 2 |
| DEAE-cellulose eluted with water (CeNPN) | 75 | ... | ... | 32 | 57 | 11 | ... | ... |
| DEAE-cellulose eluted with 0.1 M (NH$_4$)$_2$CO$_3$ | 21 | ... | ... | 54 | 26 | 9 | 7 | 4 |
| Cetavlon—non-precipitable, water-insoluble | 35 | ... | ... | 25 | 50 | 20 | ... | 5 |

[a]Quantified by integration of the peak area in g.l.c. of the alditol acetates; GalA by the carbazole method.
Reprinted from Reference [10] with permission from Elsevier Science.

of four oligosaccharides isolated from the partial hydrolysis indicated that all of the arabinose and galactose and some of the xylose were present in the side chains (Table 2.9)

The methylation analysis of carboxyl-reduced CePA indicated that the majority of the L-rhamnosyl residues were (1→2)-linked and ~46% were substituted at O-3 (Table 2.8). (1→4)-linked galacturonic acid was present essentially in the backbone chain. All of the fucose and ~50% of the galactose constituted the non-reducing end groups (Table 2.8). Partial hydrolysis of CePA with 0.25 M CF$_3$COOH yielded four oligosaccharides and significant quantities of fucose and galactose, confirming their presence in the side chains (Table 2.9).

Cui and coworkers fractionated flaxseed gum into a neutral fraction (NFG) and an acidic fraction (AFG) by applying the gum solutions directly onto a DEAE cellulose column [22]. The neutral fraction was eluted with sodium acetate buffer (pH 5.0, 25 mM), while the acidic fraction was eluted with 1.0 M NaCl in the same buffer. The relative monosaccharide compositions of the original gum and its two fractions are presented in Table 2.10. NFG, the neutral fraction, contained mainly xylose (62.8%), arabinose (16.2%), glucose

TABLE 2.8. **Methylation Analysis Data for CeNPN and Carboxyl-Reduced CePA.**

| Methyl Ether | Molar Ratio[a] | | Mode of Linkage |
|---|---|---|---|
| CeNPN | Native | Degraded | |
| 2,3,5-Me$_3$-Ara | 0·74 | ... | Ara$f$-(1→ |
| 2,3,4-Me$_3$-Xyl[b] | 7·26 | 1.20 | Xyl$p$-(1→ |
| 2,3,4-Me$_3$-Ara[b] | 8·02 | ... | Ara$p$-(1→ |
| 2,3,4,6-Me$_4$-Gal | 2·87 | ... | Gal$p$-(1→ |
| 2,4-Me$_2$-Ara | 1·51 | ... | →3)-Ara$p$-(1→ |
| 2,3-Me$_2$-Xyl | 10·58 | 4.44 | →4)-Xyl$p$-(1→ |
| 3-Me-Xyl | 1·00 | 1.00 | →2,4)-Xyl$p$-(1→ |
| Arabinose | 3·52 | ... | →2,3,4)-Ara$p$-(1→ or —2,3,5)-Ara$f$-(1→ |
| Xylose | 2·55 | ... | →2,3,4)-Xyl$p$-(1→ |
| | | Caboxyl-Reduced | |
| CePA | | CePA | |
| 2,3,4-Me$_3$-Fuc | | 2.32 | Fuc$p$-(1→ |
| 3,4-Me$_2$-Rha | | 2.40 | →2)-Rha$p$-(1→ |
| 2,3,4,6-Me$_4$-Gal | | 1.20 | Gal$p$-(1→ |
| 4-O-Me-Rha | | 1.80 | →2,3)-Rha$p$-(1→ |
| 2,3,6-Me$_3$-Gal | | 3.00 | →4)-Gal$p$-(1→ |
| Galactose | | 1.00 | →2,3,4,6)-Gal$p$-(1→ |

[a]With respect to 3-Me-Xyl for CeNPN and Gal for CePA.
[b]Quantified by g.l.c. of the alditol acetate derivatives.
Reprinted from Reference [10] with permission from Elsevier Science.

(13.6%) and galactose (7.4%). In contrast, the acidic fraction AFG contained mainly rhamnose (54.5%), galactose (23.4%) and fucose (10.1%) with only small amounts of xylose (5.5%), arabinose (2.0%) and glucose (4.5%). The presence of xylose and arabinose in AFG might originate from contamination with the neutral fraction.

Methylation analysis revealed that 71% of the xylosyl residues in NFG were from the 1,4-linked β-D-xylopyranose backbone chain while about 29% of them were from the side chains (Table 2.11). Of the backbone chain xylosyl residues, 55% are non-substituted, 10% are O-3 substituted and 35% are double substituted (2 and 3 positions). About 6.4% arabinose existed as terminal arabinofuranosyl residues, while 4.9% existed as 1,3-linked. The galactose present in NFG is only terminal residue. The small amount of 1,4-linked β-D-glucose is possibly caused by some soluble cellulose-like material from the cell wall. These results are in agreement with Muralikrishna et al.'s data that the neutral arabinoxylan fraction contained a (1→4) β-D-xylan backbone to which arabinose and galactose side chains were attached at positions 2 and/or 3 [10,26]. The methylation analysis of AFG revealed that it was mainly composed of 1,2-linked (15.4%) and 1,2,3-linked (13.1%) rhamnosyl residues with 1,4-linked

TABLE 2.9. Characterization of Oligosaccharides Obtained after Partial Hydrolysis of CeNPN and CePA.

| Oligosaccharides | Yield (%) | $R_F$ | Composition (Mole Proportion) | GalA (%) | Reducing End | $[\alpha]_D$ Water (Degrees) | DP |
|---|---|---|---|---|---|---|---|
| CeNPN | | $R_{Ma}$[a] | | | | | |
| I | 1.5 | 0.75 | Ara-Gal (1:1) | | Ara | 29 | 2 |
| II | 1 | 0.35 | Ara-Xyl (1:1) | | Xyl | 4 | 3 |
| III | 0.25 | 0.57 | Ara | | Ara | 80 | 3 |
| IV | 0.2 | 0.11 | Ara-Xyl (1:1) | | Xyl | 40 | 4 |
| CePA | | $R_{GalA}$[b] | | | | | |
| I | 5.04 | 0.84 | Rha-GalA (1:1) | 48 | Rha | 112 | 2 |
| II | 4.13 | 0.21 | Rha-GalA (2:1) | 33 | Rha | 104 | 3 |
| III | 1.35 | 0.14 | Rha-GalA (3:1) | 24 | Rha | 60 | 4 |
| IV | 0.63 | 0.1 | Rha-GalA (3:2) | 40 | Rha | 175 | 5 |

[a]Mobility in p.c. with respect to that of maltose, using 1-propanol-ethanol-water (7:1:2).
[b]Mobility in p.c. with respect to that of GalA in ethyl acetate-formic acid-acetic acid-water (18:1:3:4).
Reprinted from Reference [10] with permission from Elsevier Science.

**TABLE 2.10.** Relative Neutral Monosaccharide Compositions of Flaxseed Gum Fractions

| Sugar | CFG[a] | DFG[b] | NFG[c] | AFG[d] |
|-------|--------|--------|--------|--------|
| Rhamnose | 34.2 ± 2.5 | 35.5 ± 0.2 | 0 | 54.5 ± 0.2 |
| Fucose | 4.50 ± 0.1 | 5.00 ± 0.0 | 0 | 10.1 ± 0.7 |
| Arabinose | 9.80 ± 0.5 | 8.80 ± 0.2 | 16.2 ± 0.2 | 2.00 ± 0.0 |
| Xylose | 32.0 ± 0.5 | 29.8 ± 0.5 | 62.8 ± 5.2 | 5.50 ± 1.2 |
| Galactose | 17.3 ± 1.4 | 20.9 ± 0.0 | 7.40 ± 0.0 | 23.4 ± 1.5 |
| Glucose | 2.20 ± 0.4 | 0.90 ± 0.4 | 13.6 ± 1.1 | 4.50 ± 0.0 |

[a]CFG crude flaxseed gum.
[b]DFG dialyzed flaxseed gum.
[c]NFG neutral fraction of flaxseed gum.
[d]AFG acidic fraction of flaxseed gum.
Reprinted from Reference [22] with permission from the *Journal of Agricultural and Food Chemistry.*
© 1994 The American Chemical Society.


**TABLE 2.11.** Molar Ratio[a] of Partially Methylated Acetyl Alditols (PMAA) of Neutral (NFG) and Acidic (AFG) Fractions of Flaxseed Gums.

| PMAA | NFG | AFG | Deduced Linkage |
|------|-----|-----|-----------------|
| 2,3,5-Me$_3$-Ara | 6·4 | 0 | Terminal Ara (f)[b] |
| 2,5-Me$_2$-Ara | 4·9 | 0 | 1,3-Linked Ara (f) |
| Total ether of Ara | 11·3 | 0 | |
| 2,3,4-Me$_3$-Xyl | 16·2 | 0 | Terminal Xyl (p)[c] |
| 2,3-Me$_2$-Xyl | 21·8 | 0 | 1,4-Linked Xyl (p) |
| 2-Me-Xyl | 4·1 | 0 | 1,3,4-Linked Xyl (p) |
| Xyl(acetyl)$_5$ | 14·1 | 0 | 1,2,3,4-Linked Xyl (p) |
| Total ether of Xyl | 56·2 | 0 | |
| 2,3,4,6-Me$_4$-Gal | 0 | 17.0 | Terminal Gal (p) |
| 2,3,6-Me$_3$-Glc | 3·8 | tr[d] | 1,4-Linked Glc (p) |
| 3,4-Me$_2$-Rham | 0 | 15.4 | 1,2-Linked Rham |
| 4-Me-Rham | 0 | 13.1 | 1,2,3-Linked Rham |
| Total ether of Rham | 0 | 28.5 | |
| 2,3-Me$_2$-Gal6D$_2$ | 0 | 10.2 | 1,4-Linked-D-galacturonic acid |

[a]Molar ratio was calculated from peak area of GLC.
[b]Furanosyl ring.
[c]Pyranosyl ring.
[d]tr, trace amount.
Reprinted from Reference [22] with permission from the *Journal of Agricultural and Food Chemistry.*
© 1994 The American Chemical Society.

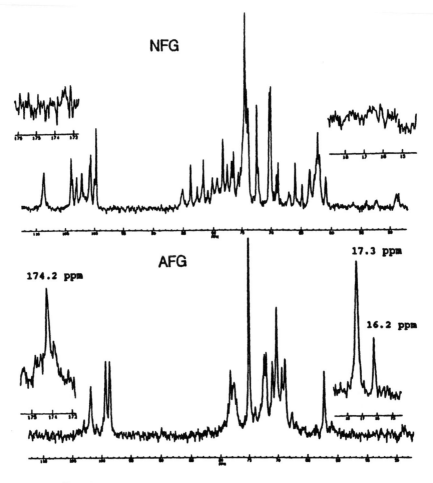

**Figure 2.7** $^{13}$C NMR spectra of neutral (NFG) and acidic (AFG) fractions of flaxseed gum (2.0% in D$_2$O (w/w), 65°C). Reprinted from Reference [22] with permission from the *Journal Agricultural and Food Chemistry.* © 1994 The American Chemical Society.

galacturonic acid (10.2%). This is also in general agreement with Muralikrishna et al.'s work except for the presence of 1,2,3-linked rhamnose [26]. The chemical structures of NFG and AFG are further confirmed by $^{13}$C NMR spectra as presented in Figure 2.7. The anomeric resonances at 108 ppm of NFG are attributed to the α-arabinose, while the resonances between 99 to 104 ppm arise from the xylose residues [26]. The lack of resonances at chemical shift between 170–180 ppm and 15 to 20 ppm of the $^{13}$C NMR spectrum of NFG supports the conclusion obtained from monosaccharide and methylation analyses that there is no acidic component in NFG. The $^{13}$C NMR spectrum of AFG is rather simple at the anomeric region, and it agrees with the results of methylation analysis. Resonance at 174.2 ppm is derived from the 1,4-linked galacturonic acid, while resonances at 17.3 ppm and 16.2 ppm are from the rhamnosyl and fucosyl residues, respectively [26].

## 3.3. MOLECULAR WEIGHT DISTRIBUTION OF FLAXSEED GUM

Gel filtration chromatographic profiles of flaxseed gum and its fractions are shown in Figure 2.8 [22]. Crude gum (CFG) and dialyzed gum (DFG) consisted mainly of high molecular species, eluting earlier than the linear dextran T-500. Substantial amounts of uronic acid were detected in CFG and DFG. The peak elution volume of the acidic species appeared later than that of neutral polysaccharides indicating that acidic polysaccharides had a smaller hydrodynamic volume. The dialysis process did not change the molecular weight and its distribution of flaxseed gum since both CFG and DFG exhibited similar gel filtration profiles, although a slight shift of the void volume is observed between CFG and DFG. As shown in Figure 2.8, the gel filtration profile of the neutral fraction (NFG) is similar to the profiles of DFG and CFG, except for the lack of the acidic element. These results are in accordance with the monosaccharide and methylation analyses and $^{13}$C NMR data that show that NFG is a pure neutral polysaccharide. The neutral carbohydrate peak observed near the void volume also suggests that NFG mainly contains high molecular size species, although smaller amounts of neutral carbohydrates are detected in the low molecular weight region. The molecular weight distribution of AFG (Figure 2.8) demonstrated that the hydrodynamic volume of AFG is much smaller than that of NFG, DFG and CFG. The molecular weights of flaxseed gum and its fraction estimated by gel filtration chromatography are in agreement with their intrinsic viscosity values, as shown in Table 2.12 [26].

## 4. PHYSICAL PROPERTIES OF FLAXSEED GUM

An early description of the functional properties of flaxseed gum by BeMiller and coworkers [27] contained information published before 1970. Recent revisits to the functional properties of flaxseed gum started in the late 1980s and early 1990s. Since then, a substantial amount of information on the functional properties, especially the rheological properties of flaxseed gum and applications, has been generated [12–14,26,28]. Earlier studies on the rheological properties of flaxseed gum appear not to be consistent with recently published data, possibly due to limitations of instrumentation and technology [27]. Therefore, the functional properties of flaxseed gum discussed in this chapter are primarily from recent publications.

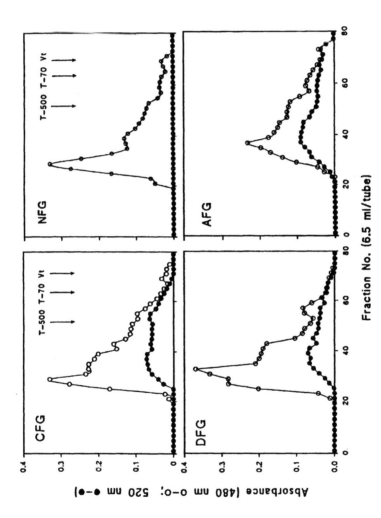

**Figure 2.8** Gel filtration chromatographic profiles of flaxseed gum: crude (CFG), dialyzed (DFG) and neutral (NFG) and acidic (AFG) fractions on Sepharose 2B column (2.5 × 90 cm, flow rate 30 mL/h at 25°C) eluted with 0.1 M NaCl in distilled water (480 nm, carbohydrates; 520 nm, uronic acids). Reprinted from Reference [22] with permission from the *Journal Agricultural and Food Chemistry*. © 1994 The American Chemical Society.

TABLE 2.12. **Comparison of Intrinsic Viscosity of Flaxseed and Commercial Gums.**

| Gum | Intrinsic Viscosity (mL g$^{-1}$, in 1 M NaCl)[a] |
|---|---|
| Arabic | 14.1 |
| Guar | 1135.4 |
| Xanthan | 1355.1 |
| Norman | |
|   CFG[b] | 483.0 |
|   NFG | 530.4 |
|   AFG | 248.4 |
| 84495 | 657.8 |
| Foster | 434.0 |
| Omega | 536.6 |
| Linola 947 | 476.3 |

[a]Based on as is weight.
[b]CFG: crude flaxseed gum; NFG: neutral flaxseed gum; AFG: acidic flaxseed gum.
Reprinted from Reference [26] with permission from Elsevier Science.

## 4.1. RHEOLOGY: VISCOSITY AND FLOW BEHAVIOR OF FLAXSEED GUM

Flaxseed gum exhibits Newtonian flow behavior at low concentrations and shear thinning flow behavior at high concentrations, as demonstrated in Figure 2.9. The measured viscosity is independent of shear rate in a Newtonian solution; in contrast, in shear thinning flow behavior, the viscosity decreases as the shear rate increases. Gums extracted from different flax varieties exhibit wide diversities in chemical compositions and structures, as well as rheological behaviors. For example, Wannerberger and coworkers examined 23 flax varieties and found that most of the gums at concentrations ≥ 1.0% exhibited shear thinning flow behaviors; whereas, a Newtonian flow behavior was observed for flaxseed gum from cv. Liflora [12]. Similar variations with varieties were also observed by Cui and coworkers in comparison of gums from six yellow and six brown varieties [21]. It appeared that the variation of flow behavior is correlated with the structure features of flaxseed gum: gums that contain high levels of neutral polysaccharides tend to exhibit shear thinning flow behavior, and vice versa.

Since the flow behaviors of flaxseed gums demonstrate significant dependence on their chemical structures, a gum extracted from cv. Norman is fractionated into a neutral flaxseed gum (NFG) and an acidic gum (AFG), as described in section 3. The flow behaviors of the NFG and AFG are compared to their original material, CFG and DFG, as shown in Figure 2.10. CFG exhibits shear thinning behavior at polymer concentrations above 1.0% over a broad range of

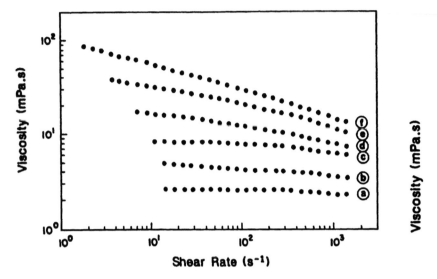

**Figure 2.9** Flow curves of flax seed gum (pH 6.5, temp 25°C) at 0.05% (a), 0.1% (b), 0.2% (c), 0.3% (d), 0.4% (e), and 0.5% (f). Reprinted from Reference [13] with permission from IFT.

shear rates (0.1 to 1000 $s^{-1}$). A slight shear thinning behavior for this material is observed at a high shear rate region when polymer concentration is 0.5% or less. Similar shear thinning flow behavior is also observed for DFG. NFG in solutions exhibits shear thinning flow behavior when polymer concentrations are above 0.5%. At 0.3%, the steady shear flow curve of this fraction is Newtonian-like. The shear thinning behavior of NFG could be attributed to its higher molecular weight arabinoxylan component (Figure 2.8). In contrast, the flow behavior of AFG solutions are typical of Newtonian flow at all concentrations examined. These rather weak rheological responses of AFG are ascribed to its much smaller molecular weight and low intrinsic viscosity (Figure 2.8 and Table 2.12). The overall ranking of viscosity is in the order of CFG > DFG > NFG >AFG [22].

## 4.1.1. Effect of pH

The flow behavior and viscosity of flaxseed gum are affected by solution pHs and presence of co-solutes. As shown in Figure 2.11, the lowest viscosity for flaxseed gum is observed at pH 2. As the pH increases, the viscosity increases steadily until pH 8, at which the viscosity jumps to three times its value at pH 2. A further increase of pH resulted in a decrease of viscosity [13]. This observation is in accordance with the result of Susheelamma [11]. A later study by Fedeniuk and Biliaderis showed similar trends at a lower pH region, however, the highest viscosity was observed between pH 5–10 and a further increase in

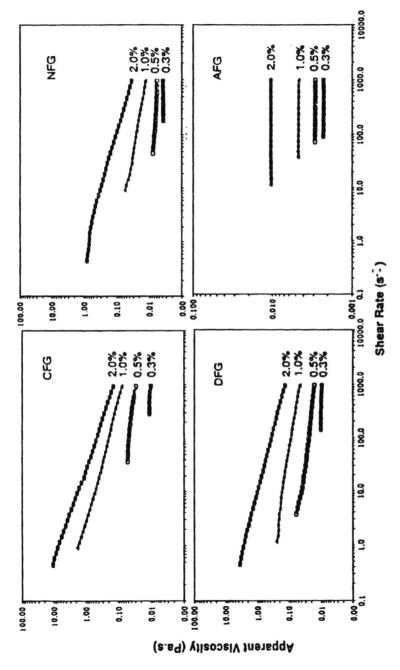

**Figure 2.10** Steady shear rheological flow curves of crude (CFG), dialyzed (DFG), neutral fraction (NFG) and acidic fraction (AFG) solution (pH 6.5) of flaxseed gum at 25°C. Reprinted from Reference [22] with permission from the *Journal Agricultural and Food Chemistry*. © 1994 The American Chemical Society.

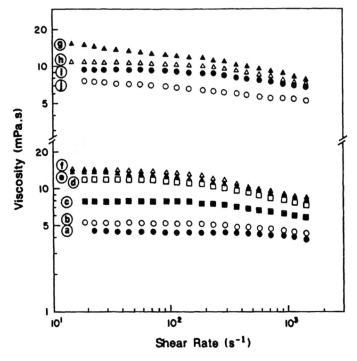

**Figure 2.11** Effect of pH on viscosity of 0.3% (w/v) mucilage solutions (25°C): pH 2.0 (a), 3.0 (b), 6.0 (e), 7.0 (f), 8.0 (g), 9.0 (h), 10.0 (i), and 12.0 (j). Reprinted from Reference [13] with permission from IFT.

pH resulted in a decrease in viscosity (Figure 2.12) [14]. The slight discrepancy of the effect of pH on viscosity of gum solutions might be due to structure and composition differences of the gums studied.

### 4.1.2. Effect of Solute

As shown in Figures 2.13 and 2.14, respectively, the addition of NaCl to flaxseed gum solutions resulted in a reduction of viscosity. This salt effect is more pronounced with an increasing electrolyte concentration up to 0.2%. A further increase of NaCl concentrations has little additional effect on the reduction of viscosity. The addition of NaCl might increase the association of counter-ions with the polymer molecules which consequently might suppress the electrostatic repulsion of charged groups on the polymer chains. This allows the polysaccharide molecules to adopt a more compact conformation in solutions, therefore, the viscosity drops. As the NaCl concentration increases, the repulsion of the polymer molecules decreases and so does the viscosity.

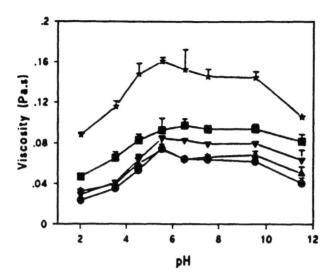

**Figure 2.12** Effect of pH on the apparent viscosity of 1.0% (w/w) flaxseed gum and fraction solutions (116s$^{-1}$, 25°C) [exception: meal 55°C, 1.5% (w/w)]: Linott 80°C (●); Linott 4°C (■); meal 55°C (▲); CP fraction (▼); CS fraction (★). Reprinted from Reference [14] with permission from the *Journal of Agricultural and Food Chemistry*. © 1994 The American Chemical Society.

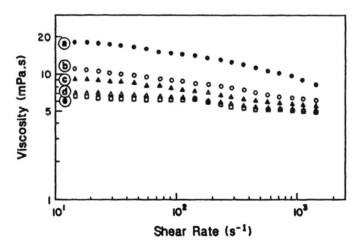

**Figure 2.13** Effect of NaCl on the viscosity of flaxseed gum (0.3% w/v, 25°C): no salt (a); NaCl 0.02% (b), 0.05% (c), 0.20% (d), and 2.00% (e). Reprinted from Reference [13] with permission from IFT.

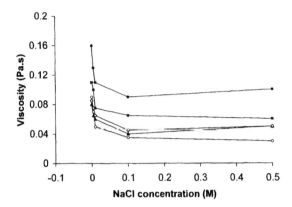

**Figure 2.14** Effect of NaCl concentration on the apparent viscosity of 1.0% (w/w) gum and fraction solutions ($116s^{-1}$, 25°C) [exception: Meal 55°C, 1.5% (w/w)]: Linott 80°C (O), Linott 4°C (■), Meal 55°C (◇); CP fraction (▲); CS fraction (●). Reprinted from Reference [14] with permission from the *Journal of Agricultural and Food Chemistry.* © 1994 The American Chemical Society.

When the concentration of NaCl is increased to an extent that the $Na^+$ counter-ion becomes saturated, further addition of NaCl would have little additional effect on the reduction of viscosity [14]. Other salts also significantly reduced the viscosity of flaxseed gum. For example, sodium and potassium chloride reduced the viscosity of a flaxseed gum by 10%, calcium chloride by 20%, ferric chloride by 50% and potassium iodide by 40% [11]. In contrast to the above observation, the presence of carbonates of calcium and magnesium increased the viscosity of flaxseed gum by 10–20%, depending on salt concentration [11]. If the salt concentration exceeds a certain level, e.g., 0.5 M for carbonates, aggregation occurs, which essentially leads to the precipitation of the polysaccharides [11].

### 4.1.3. Effect of Temperature

The effect of temperature on the viscosity of flaxseed gum was reported by Susheelamma [11], Mazza and Biliaderis [13] and Fedeniuk [24]. An expected trend of decreasing viscosity with increasing temperature is observed, and it follows an Arrhenius type of dependence, as shown in Figure 2.15 [13]. Based on this experiment, energy of activation values ($E_a$) are calculated, as summarized in Table 2.13. The activation energy is determined by polymer concentration as well as by the primary structure of the gum. For example, 1% of flaxseed gum solutions from Linott extracted at 80°C and 4°C, respectively, have a similar $E_a$ value (23 kJ/mol) while 0.3% of flaxseed gum solution gives an $E_a$ of 13.3 kJ/mol (Table 2.13) [13]. As a comparison, $E_a$'s for 1% guar gum and carboxymethylcellulose are 11.7 kJ/mol and 23 kJ/mol, respectively [24,29]. A neutral polysaccharide fraction of flaxseed gum, the CTAB-soluble

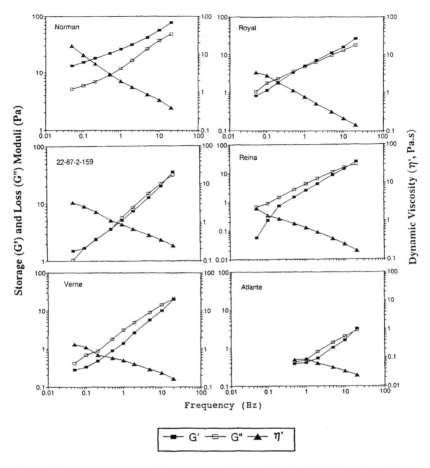

**Figure 2.15** Mechanical spectra of flaxseed gum solutions (2.0% w/w, 25°C) from selected brown flaxseed cultivars. Reprinted from Reference [21] with permission from Elsevier Science.

(CS) fraction, has a substantially higher $E_a$ (30.8 kJ/mol) than all the other gum solutions; this may imply that the neutral polymers are more closely associated with each other, thus requiring higher energy to initiate the viscous flow compared to the acidic fraction (CTAB-precipitated, CP), which has a much smaller $E_a$ value (22.8 kJ/mol) [24]. Susheelamma also reported a decrease in viscosity with increased temperature; a greater decrease of viscosity with increase of temperature occurred at higher gum concentrations [11].

## 4.2. RHEOLOGY: VISCOELASTIC PROPERTIES

Dynamic oscillatory rheological tests showed that flaxseed gums extracted from different varieties exhibited wide variations in their viscoelastic properties.

TABLE 2.13. **Activation Energy ($E_a$) of Flaxseed Gum and Its Fraction Solutions (1.0% w/w) (116 s$^{-1}$, 15–65°C, pH 5.5).**

| Flaxseed Gum Source | $E_a$ (kJ/mol)[a] |
|---|---|
| Linott 4°C | 23.5 |
| Linott 80°C | 22.8 |
| CTAB-precipitated fraction (CP) | 22.8 |
| CTAB-soluble fraction (CS) | 30.8 |

[a]Linear regression of log $\eta = $ log A $+ 2.303$ ($E_a/RT$), $r^2 > 0.98$, $p < 0.01$. Reprinted from Reference [24].

Wannerberger and coworkers reported that the solution of gum from cv. Liflora was a typical viscoelastic fluid at concentrations of 1–3%; in contrast, gum from cv. Szegedi 62 exhibited weak gel structure when compared at the same concentration [12]. The trend of elasticity of flaxseed gums is similar to that of viscosity as described in section 4.1; it follows an increased order of Liflora < Linda < Ariadna < Belinka < Szegedi 62; this order is correlated to the amount of arabinoxylan in the gum [12].

The varietal dependences of viscoelastic properties of flaxseed gums have also been demonstrated by Cui and coworkers [21]. Of 12 gums (from 12 varieties) examined, Norman gum (2.0%, w/w) exhibited weak gel properties with the storage modulus ($G'$) greater than the loss modulus ($G''$) over the entire frequency examined (Figure 2.15). Gum extracted from cv. Royal seed behaved like a typical viscoelastic fluid where the storage modulus was smaller than the loss modulus at the lower frequency range, and the reverse was observed at higher frequencies. Gums from Reina, Verne and Atlante were of viscoelastic fluid, in which the loss modulus ($G''$) was higher than the storage modulus ($G'$) over most of the frequency range investigated. Of the gums extracted from yellow flaxseed, only cv. Omega exhibited liquid-like properties (Figure 2.16). The mechanical spectra of gum solutions from cv. 84495 and 92-5103 were typical of weak gels in that the storage modulus ($G'$) was much greater than the loss modulus ($G''$) and the two moduli were less frequency dependent compared to viscoelastic fluid (Figure 2.16). The mechanical spectra of APF-9006 and APF-9007 are superimposable, indicating that gums extracted from closely related breeding lines contain polysaccharides of similar chemical structures and exhibit similar rheological properties (Figure 2.16 and Table 2.5).

Data presented in Figures 2.15 and 2.16 also show that the average dynamic viscosity of gums from selected yellow-seeded cultivars is higher than that from brown-seed gums. Of the yellow-seed gums, cultivar 84495 exhibited the highest dynamic viscosity followed by APF 9007, APF 9006, 92-5103, Foster

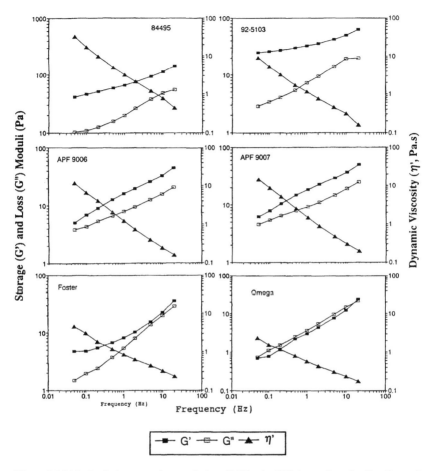

**Figure 2.16** Mechanical spectra of gum solutions (2.0% w/w, 25°) from selected yellow flaxseed cultivars. Reprinted from Reference [21] with permission from Elsevier Science.

and Omega. Of the gums from brown seed, Atlante had the lowest dynamic viscosity, Norman had the highest with all the others in between (Figure 2.15).

Phase angle is defined as tan $\delta = G''/G'$. The value of phase angle reflects the departure of a system from the solid-like status. A phase angle of 0° indicates a solid phase, while a value of 90° is a typical liquid. If a system has a phase angle smaller than 45°, the system has a more solid-like character than a liquid-like character, and vice versa. Figure 2.17 shows the phase angle values of gums from selected brown and yellow seed. Although the phase angle values changed with frequency, most of the brown-seed gums had phase angle values greater than 45°, except for Norman. In contrast, phase angle values of all of the yellow-seeded gums are smaller than 45°, except for Omega.

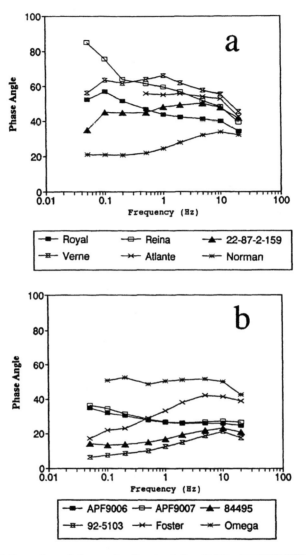

**Figure 2.17** Comparison of phase angle of gum solutions (2.0% w/w, 25°C) of brown (a) and yellow (b) flaxseed. Reprinted from Reference [21] with permission from Elsevier Science.

The dynamic rheological properties of flaxseed gum (cv. Norman) and its two fractions, AFG and NFG, are presented in Figure 2.18. Both CFG and DFG exhibited weak gel properties, however, the gel strength of DFG is much weaker than that of CFG. It is rather surprising that following dialysis, DFG exhibited lower apparent viscosity and much weaker viscoelastic responses compared to CFG (Figure 2.18). This would suggest that a small molecular weight

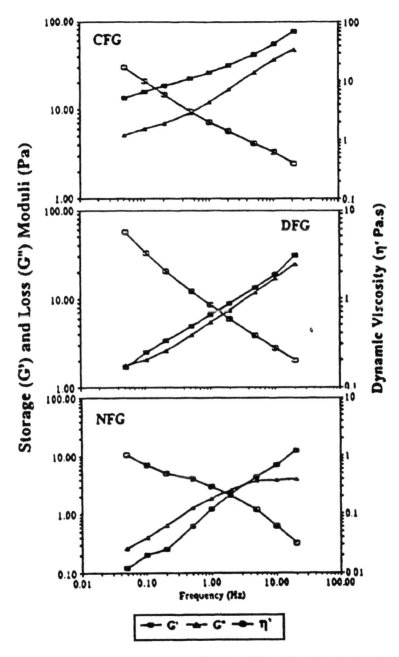

**Figure 2.18** Frequency dependence of storage ($G'$) and loss ($G''$) moduli and dynamic viscosity ($\eta'$) of 2.0% (w/w) solutions of crude (CFG), dialyzed (DFG) and neutral fraction (NFG) flaxseed gums (pH 6.5, 25°C). Reprinted from Reference [22] with permission from the *Journal of Agricultural and Food Chemistry.* © 1994 The American Chemical Society.

component (removed by dialysis) in the crude gum might be responsible for the highly viscous character of CFG. When DFG is separated into neutral and acidic fractions, the dynamic rheological responses of the two fractions are different. The dynamic rheological pattern of NFG is typical of a viscoelastic fluid similar to that of guar gum in that the $G''$ is greater than $G'$ at lower frequencies, while the reverse is observed at higher frequencies [22]. A similar dynamic spectrum was reported for a neutral fraction isolated from cv. Linott by CTAB precipitation [14]. The acidic fraction (AFG) is a much weaker viscoelastic fluid in which the $G''$ is higher than the $G'$ over the entire frequency invested (0.01 Hz to 20 Hz, data not shown).

## 4.3. WATER-HOLDING CAPACITY (WHC)

The water-holding capacity (WHC) of three flaxseed gum is compared against xanthan and guar gums, as shown in Table 2.14. The water-holding capacity of the three gums follows a decreased order of Linott 4°C, Linott 80°C and Meal 55°C, which coincides with the order of total carbohydrate content (Table 2.2). The WHC of most food hydrocolloids ranges from 300 to 3200 g of $H_2O/100$ g of solids [30]. The WHC values obtained for flaxseed gums in the range of 1600 to 3000 g of $H_2O/100$ g of solids, indicate that flaxseed gum has a relative high WHC. The equilibrium WHC values of flaxseed gums could be obtained within 3 hr; in contrast, the equilibrium WHC values of guar and xanthan gums need at least over 30 hr when using the same methods [14]. This result suggests that flaxseed gums are easily hydrated, which may have some advantages when the hydration rate of a gum is important.

TABLE 2.14. **Water-Holding Capacity (WHC) of Commercial Food Gums and Flaxseed as Determined by the Baumann Capillary Apparatus [24].**

| Material | Water-Binding Capacity[a] (g $H_2O/100$ g solids) |
|---|---|
| Xanthan gum | $32300 \pm 1100$ |
| Guar gum | $2200 \pm 400$ |
| Flaxseed gum from Linott 4°C | $3000 \pm 300$ |
| Flaxseed gum from Linott 80°C | $2500 \pm 100$ |
| Flaxseed gum from Meal 55°C | $1600 \pm 100$ |

[a]Means $\pm$ S.D. ($n = 3$).
Reprinted from Reference [24].

## 4.4. SURFACE TENSION AND THE STABILITY OF FOAMS AND EMULSIONS

### 4.4.1. Surface Tension

Flaxseed gums extracted from Linott 80°C, Linott 4°C and Meal 55°C have the ability to reduce the surface tension of water, as shown in Figure 2.19. For the two gums from Linott 80°C and Meal 55°C, the surface tension of water decreased rapidly with increasing gum concentration and then leveled off to a constant value at gum concentrations above 0.3% [24]. However, the surface tension of water decreased almost linearly with increasing concentration of gum from Linott 4°C (Figure 2.19). The observed deviation of surface activity among flaxseed gums might be attributed to their differences in protein content (Table 2.2). Proteins are considered primarily responsible for surface active properties of polysaccharide gums, although some modified polysaccharides, in particular those that have hydrophobic groups, such as methyl or acetyl groups, were also demonstrated to be surface active [31].

### 4.4.2. Foam Stability

The combination of surface activity and viscosity of polysaccharide gums exerts protective action against thermal disruption of foams. Figure 2.20 shows the effect of concentration of flaxseed gum on foam stability. In 1% (w/v) gum

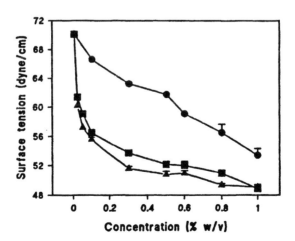

**Figure 2.19** Reduction of surface tension of water by mucilages (pH 5.5, 25°C); Linott 4°C (●), Linott 80°C (■) and Meal 55°C (▲). Reprinted from Reference [24] with permission.

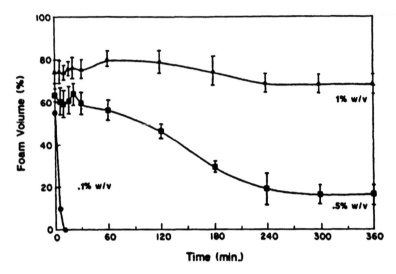

**Figure 2.20** Foam stability of flax seed mucilage solution of 0.1 (●), 0.5 (■) and 1.0 (▲) % (w/v); values are expressed as percent of the respective foam volumes of ovalbumin solutions at the same concentration. Reprinted from Reference [13] with permission from IFT.

solution, flaxseed gum gave foam values about 75% of those of ovalbumin and had similar time-dependent stability [13]. The foam values of 0.5% and 0.1% solutions were 63% and 57% of the respective ovalbumin solutions, and they were far less stable over time [13]. Using defatted oilseed flours as a source of surface active protein and $NaHCO_3$ and citric acid as an in situ source of $CO_2$, Susheelamma tested flaxseed gum at different concentrations [11]. As shown in Table 2.15, peanut and sesame proteins by themselves in the presence of a source of gas such as $CO_2$ had a tendency to form foams with a smaller increase in foam volume after heating. In the presence of a low concentration of flaxseed gum, the foam column expanded considerably but was not stable to heat. At slightly higher gum concentrations, the foam column decreased in volume due to the compactness of the foam but was stable against thermal disruption. This behavior could be explained as follows. The addition of a small amount of flaxseed gum imparted some viscosity to the aqueous solution, which in turn, exerted a beneficial effect on the surface-active protein. However, foams formed were unstable to heat [11]. At slightly higher gum concentrations, the viscosity of aqueous phase increased significantly. The increase in viscosity partly minimizes the free expansion of the foam columns and also makes them more compacted, and this resulted in a decrease in foam volume but with increased thermal stability. At higher gum concentrations (1.5% and above), the foam columns are stable even at high temperatures. A neutral flaxseed gum fraction, which exhibited the highest viscosity, stabilized foams most effectively during heating [24]. These results confirmed the rule of polysaccharide gums

TABLE 2.15. The Effect of Flaxseed Gum and Guar Gum on the Foam Formed by Peanut and Sesame Flours.

| Component Wt (mg) | | | | Foam Vol[a] (mL) | | | |
|---|---|---|---|---|---|---|---|
| Linseed Mucilage | Guar Gum | Peanut Flour | Sesame Flour | After Acidification | After 10 min of Acidification | At 95°C | After Cooling |
| 0 | | 30 | | 2 | 2 | 5 | 0.5 |
| 0 | | | 30 | 3 | 2 | 6 | 0.5 |
| 5 | | 30 | | 6 | 5 | 15 | 1.0 |
| 10 | | 30 | | 6 | 4 | 16 | 1.0 |
| 15 | | 30 | | 5 | 3 | 15 | 4.0 |
| 20 | | 30 | | 4 | 2 | 12 | 3.0 |
| 5 | | | 30 | 6 | 5 | 16 | 1.0 |
| 10 | | | 30 | 5 | 4 | 18 | 1.0 |
| 15 | | | 30 | 5 | 4 | 14 | 4.0 |
| 20 | | | 30 | 5 | 4 | 16 | 4.0 |
| | 5 | 30 | | 7 | 6 | 10 | 1.0 |
| | 10 | 30 | | 6 | 5 | 10 | 1.0 |
| | 15 | 30 | | 4 | 4 | 12 | 3.5 |
| | 20 | 30 | | 3 | 3 | 14 | 3.0 |
| | 5 | | 30 | 5 | 4 | 17 | 1.0 |
| | 10 | | 30 | 5 | 4 | 18 | 1.0 |
| | 15 | | 30 | 4 | 3 | 18 | 5.0 |
| | 20 | | 30 | 4 | 3 | 16 | 4.0 |

[a]The volume before acidification was 1.75 mL in all cases.
Reprinted from Reference [11] with permission from IFT.

93

in the stabilization of foams: the initial formation of foam is impeded by the addition of gum due to an increase in viscosity of the liquid medium. The viscosity and elasticity of the thin film surrounding the gas bubbles are important for foam stability. The increased viscosity on one hand slows gas diffusion and, on the other, stabilizes the films surrounding the gas bubbles.

### 4.4.3. Emulsion Stability

As shown in Table 2.16, an increase in the concentration of flaxseed gum from Linott 80°C and Meal 55°C in the continuous phase facilitated the formation of emulsion as evidenced by an increasing interfacial area of the emulsion and centrifugal stability. In contrast, the incorporation of gum of Linott 4°C into the aqueous phase initially increased the emulsion quality, however, a higher gum concentration hindered the emulsion formation, which is similar to that of xanthan gum [24]. Because the rheological properties of gums of Linott 4°C

TABLE 2.16. Centrifugal Stability and Interfacial Area Data for Oil/Water Emulations Prepared Using Commercial and Flaxseed Gums [24].

| Polymer | Interfacial Area ($10^{-4}$ cm$^{-1}$) | Water Separation (mL) | Cream Layer (mL) | Oil Separation (mL) |
|---|---|---|---|---|
| Distilled water | 0·16 | 8·1 | 1·2 | 1·3 |
| Gum arabic 0.1% | 0·16 | 8·4 | 0·5 | 1·6 |
| Gum arabic 0.3% | 0·22 | 7·7 | 1·7 | 1 |
| Gum arabic 0.5% | 0·26 | 7·5 | 2·7 | 0·3 |
| Gum arabic 1.0% | 0·35 | 6·4 | 4 | 0 |
| Gum arabic 1.5% | 0·43 | 6·1 | 4·4 | 0 |
| Gum arabic 2.0% | 0·51 | 3·7 | 6·8 | 0 |
| Xanthan 0.1% | 0·26 | 7·6 | 2 | 0·9 |
| Xanthan 0.3% | 0·21 | 6·3 | 4·1 | 0·1 |
| Xanthan 0.5% | 0·11 | 6·7 | 3·7 | 0·2 |
| Linott 80°C 0.1% | 0·08 | 8·5 | 0·9 | 1·2 |
| Linott 80°C 0.3% | 0·1 | 8·2 | 2·2 | 0·1 |
| Linott 80°C 0.5% | 0·2 | 7·5 | 3 | 0 |
| Linott 80°C 1.0% | 0·32 | 6·8 | 3·7 | 0 |
| Linott 4°C 0.1% | 0·13 | 7·8 | 2·1 | 0·6 |
| Linott 4°C 0.3% | 0·16 | 7·4 | 2·9 | 0·1 |
| Linott 4°C 0.5% | 0·16 | 7·1 | 3·4 | 0·1 |
| Linott 4°C 1.0% | 0·13 | 7·9 | 2·5 | 0·1 |
| Meal 55°C 0.1% | 0·09 | 8·8 | 1·1 | 0·5 |
| Meal 55°C 0.3% | 0·19 | 7·5 | 2·9 | 0·1 |
| Meal 55°C 0.5% | 0·19 | 7·4 | 2·9 | 0·1 |
| Meal 55°C 1.0% | 0·25 | 6·6 | 3·8 | 0 |
| Meal 55°C 1.5% | 0·25 | 5·4 | 4·9 | 0 |

and Linott 80°C are similar (section 4.2), the differences in emulsion properties could be attributed to their differences in protein content, which might affect the surface and interfacial activities of flaxseed gum [24].

# 5. FOOD AND NONFOOD APPLICATIONS OF FLAXSEED GUM

Flaxseed gum has not been used to any degree as an industrial gum due to limited information on functional properties and availability. Recent research advances generated a great deal of new knowledge regarding the functional properties of flaxseed gum, which in turn, stimulated more studies on the application of flaxseed gum in food and nonfood products.

## 5.1. APPLICATIONS OF FLAXSEED GUM IN FOODS

Similar to other gums, flaxseed gum can be used as a stabilizer to improve food quality. The effect of flaxseed gum on bread-making properties, including pasting, dough rheology and baking, was carried out by Garden-Robinson [32]. The addition of flaxseed gum did not change the pasting temperature, however, 0.5% and 1% crude flaxseed gum significantly decreased the peak height (Table 2.17), possibly due to competition for hydration between the starch and added gum. The effect of flaxseed gum on dough rheology is reflected by farinograph properties. As shown in Table 2.18, the absorption increased by approximately 2% for every 0.5% flaxseed gum added. Choosing wheat flour with a higher absorption can increase bread yield and, consequently, profits. Farinograph peak times also significantly increased with the addition of flaxseed gum, however, 1% of flaxseed gum significantly weakened the farinogram curve [32]. The extensigraph studies showed that extensibility and resistance to extension

TABLE 2.17. **Properties of Wheat Starch with Added Crude Flaxseed Gum (CFG) or Xanthan Gum (XG).**[a]

| Sample | Pasting Temp. (C) | Peak Ht (BU) | 15 Min. Ht (BU) | Setback (BU) |
|---|---|---|---|---|
| Wheat starch (*K*) | 62·5 a | 630 b | 735 c | 850 c |
| *K* + 0.5% CFG | 62·5 a | 250 d | 520 d | 835 c |
| *K* + 1.0% CFG | 62·5 a | 390 c | 725 c | 1300 a |
| *K* + 0.5% XG | 59·5 b | 600 b | 860 b | 975 b |
| *K* + 1.0% XG | 59·5 b | 1170 a | 1295 a | 1325 a |

[a]Means followed by the same letter in columns are not significantly different at $p = 0.05$ according to Duncan's New Multiple Range Test ($n = 2$).
Reprinted from Reference [32].

**TABLE 2.18. The Effects of Crude Flaxseed Gum (CFG), Purified Flaxseed Gum (PFG) and Xanthan Gum (XG) on Farinograph Properties of Len Hard Red Spring Wheat Flour ($K$).[a]**

| Sample | Absorption[b] (%) | Peak Time (min) | MTI[c] (BU) | Stability (min) |
|---|---|---|---|---|
| Control ($K$) | 60·5 h | 8·3 f | 10 f | 21·8 d |
| $K$ + 0.5% CFG | 62·6 g | 10·8 e | 25 c, d | 20·3 e |
| $K$ + 1.0% CFG | 64·9 b | 11·4 e | 27.5 b, c | 18·1 f |
| $K$ + 2.0% CFG | 68·6 b | 17·4 b | 32.5 a | 10·5 g |
| $K$ + 0.5% PFG | 63·0 f | 10·8 e | 22.5 d | 19·6 e |
| $K$ + 1.0% PFG | 65·2 d | 11·1 e | 30 a, b | 19·3 e, f |
| $K$ + 2.0% PFG | 68·7 b | 16·1 c | 30 a, b | 10·0 g |
| $K$ + 0.5% XG | 63·5 e | 12·6 d | 10 f | 32·0 c |
| $K$ + 1.0% XG | 66·0 c | 16·1 c | 10 f | 34·0 b |
| $K$ + 2.0% XG | 69·3 a | 25·8 a | 15 e | 38·0 a |

[a]Means followed by the same letter in columns are not significantly different at $p = 0.05$ according to Duncan's New Multiple Range Test ($n = 2$).
[b]Calculation is based on a 14.0% moisture basis.
[c]Mixing tolerance index.
Reprinted from Reference [32].

decreased in samples containing flaxseed gum. The proportional number, i.e., the resistance divided by the extensibility, remained constant for all levels of flaxseed gum addition, indicating a uniform mellowing of the gluten [32]. In baking studies, the mixing time, as measured by the time required to form a thin smooth film when the dough is stretched, decreased slightly with the addition of 0.5% flaxseed gum (Table 2.19). Adding flaxseed gum to bread formulations improved the grain and texture of the bread loaves. The cells were more

**TABLE 2.19. The Effects of Crude Flaxseed Gum (CFG) and Xanthan Gum (XG) Addition on the Baking Properties of Len Hard Red Spring Wheat Flour ($K$).[a]**

| Sample | Absorption[b] (%) | Mixing Time (min) | Ovenspring (cm)[c] | Specific Volume (cc/g) |
|---|---|---|---|---|
| Control ($K$) | 60.5 | 5.00 | 3.0 c | 7.0 e |
| $K$ + 0.5% CFG | 61 | 4.75 | 3.6 b | 7.7 b |
| $K$ + 1.0% CFG | 63 | 5.00 | 3.8 a | 8.1 a |
| $K$ + 0.5% XG | 61.5 | 5.75 | 3.8 a | 7.5 c |
| $K$ + 1.0% XG | 64 | 6.00 | 3.7 b | 7.3 d |

[a]Means followed by the same letter in columns are not significantly different at $p = 0.054$ according to Duncan's New Multiple Range Test ($n = 2$).
[b]Calculation is based on a 14.0% moisture basis.
[c]Calculated as baked height (cm)-proof height (cm).
Reprinted from Reference [32].

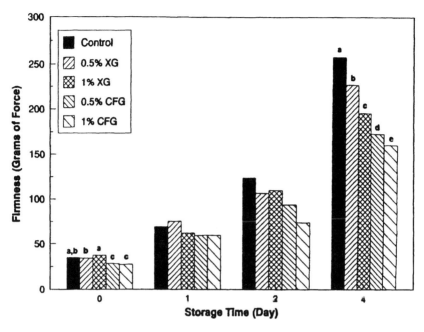

**Figure 2.21** The effects of crude flaxseed gum (CFG) and xanthan gum (XG) addition on the firmness of bread baked from Len hard red spring wheat flour. Reprinted from Reference [32].

elongated, and the texture was silky. Bread containing flaxseed gum is also significantly softer and less firm than the control at the beginning of the storage. As storage time increased, firmness values increased in all treatments. However, bread with flaxseed gum exhibited the highest softness (least firmness) after four days of storage, and the bread with 1% gum is softer than that of 0.5% gum, as shown in Figure 2.21. The structure of flaxseed gum, with its arabinoxylan component, may have played a role in its ability to delay the firming. The gum may have a mellowing effect on the gluten, which allows for greater expansion during the fermentation and baking stages. The addition of flaxseed gum was also effective in improving the subjective characteristics, ovenspring and specific volume of lower protein commercial flour [32].

Applications of flaxseed gum in other food systems were recently reported by Stewart [28]. The addition of flaxseed gum improved muffin height and volume but did not change the texture significantly, as measured by instrumental and sensory tests (Table 2.20). It was found that 0.5% flaxseed gum (flour basis) could replace 0.1% of xanthan gum or guar gum to produce muffins with good volume and texture; however, no significant increase in moisture retention or in the retard of staling were observed for all gums in the muffin study [28]. In evaluating the effect of flaxseed gum on the stability of a model salad dressing,

TABLE 2.20. **Effect of Gum Type and Concentration on Physical Characteristics of Muffins.**

| Treatment | Volume[a] (cm$^3$) | Height[1] (mm) | Weight[1] (g) |
|---|---|---|---|
| Control | 92.49[b] | 40.0[c] | 45.26[d] |
| 0.5% Flaxseed gum | 98.53[d] | 42.8[d] | 45.11[d] |
| 0.1% Flaxseed gum | 94.17[c] | 40.7[c] | 45.25[d] |
| 0.1% Xanthan gum | 98.15[d] | 41.7[acd] | 45.27[d] |
| 0.1% Guar gum | 98.04[d] | 39.8[o] | 45.25[d] |

[a]Mean of five muffins per batch × four batches.
[b,c,d]Values in the same column bearing the same superscript are not significantly different ($p < 0.05$) according to Tukey's Studentized Range test.
Reprinted from Reference [28].

a response surface methodology was successfully used to identify the effects of gum concentration, pH and salt concentration on the stability. The influence of independent variables (pH, salt and gum concentration) on the mean droplet size and rate of coalescence is demonstrated in Table 2.21. Of the independent variables examined, pH affected the stability of the emulsion the most. The most stable emulsion occurred at pH 6, while the least stable occurred at pH 2; this observation correlates with the viscosity dependence on pH: flaxseed gum exhibits the highest viscosity at pH 6 and the lowest viscosity at pH 2, as described in section 4.2. Flaxseed gum appears to act as a steric stabilizer, which means sufficient adsorbing of flaxseed gum is required to cover the particle surface completely to prevent two particles from approaching one another [28].

TABLE 2.21. **Influence of the Independent Variables (pH; Salt and Gum Concentration) on the Mean Droplet Size and the Rate of Drop Coalescence of Oil/Water Emulsion.**

| Formulation | Droplet Size Day 0 (μm) | Droplet Size Day 40 (μm) | Rate of Drop Coalescence (s$^{-1}$ × 10$^{-7}$) |
|---|---|---|---|
| 0% Gum; 2.5% salt; pH 4 | 1.31 | 1.78 | 2.69 |
| 1.5% Gum; 2.5% salt; pH 4 | 0.88 | 1.34 | 3.68 |
| 0.75% Gum; 2.5% salt; pH 4 | 0.92* | 0.97* | 0.44* |
| 0.75% Gum; 0% salt; pH 4 | 0.98 | 1.10 | 1.00 |
| 0.75% Gum; 5.0% salt; pH 4 | 0.90 | 1.03 | 1.16 |
| 0.75% Gum; 2.5% salt; pH 2 | 3.53 | 3.70 | 0.41 |
| 0.75% Gum; 2.5% salt; pH 6 | 0.79 | 1.13 | 3.13 |

*Mean of the center points.
Reprinted from Reference [28].

## 5.2. APPLICATIONS OF FLAXSEED GUM IN NONFOOD SYSTEMS

Earlier applications of flaxseed gum were summarized by BeMiller and coworker [27]. It has been suggested that flaxseed gum could be used in medicinal preparations: ointments and pastes containing flaxseed gum are effective in the treatments of furunculosis, carbunculosis, impetigo, and ecthyma [33]. Flaxseed gum could be used as a bulk laxative, a cough emollient agent and a stabilizer in barium sulphate suspensions for X-ray diagnostic preparations [34,35]. Tablets prepared with flaxseed gum have improved disintegration and allowed for a slower rate of drug release [36]. The stringy and fast drying properties of flaxseed gum make it suitable in hairdressing preparations [37], hand cream formulations [38] and denture adhesives [39]. Flaxseed gum is also considered to be a better water-in-oil emulsifier than Tween 80, gum arabic or gum tragacanth [40]. Concentrations of 0.5–1.5% flaxseed gum are suitable to stabilize oil/water emulsions. At 2.5% concentration, flaxseed gum is a good base for an eye ointment. In food applications, flaxseed gum has been used as an egg white substitute in bakery products and ice cream [41]. The strong buffering action of flaxseed gum also makes it useful in the manufacture of fruit drinks [42]. Other applications of flaxseed gum include in printing, textile and cigar industries [17]. A paper product using flaxseed gum as a deflocculant has good tensile and flexural (tear) strength [43].

Recently, flaxseed gum solution has been used as a saliva substitute because it possesses a very unusual combination of rheological and surface-chemical properties, e.g., lubricating and moisture-retaining characteristics resembling those of natural saliva [44]. The suitable viscosity of flaxseed gum as a saliva substitute is from 1–30 mPa.s, preferably 2–10 mPa.s. The product is reported to be effective for patients suffering from dryness of the mouth (xerestomia) and particularly suitable for use in dryness of the mouth at night [44].

## 6. REFERENCES

1. Pryde E. *Non-Food Used Vegetable Oils. HandBook of Processing and Utilization and Agriculture,* Volume 2. Wolff, ed. Boca Raton, FL: CRC Press Inc.; 1983. p. 109.

2. Carter J. F. Potential of flaxseed and flaxseed oil in baked goods and other products in human-nutrition. *Cereal Foods World* 1993; 38:753–9.

3. Oomah B. D., Mazza G. Flaxseed proteins—A review. *Food Chemistry* 1993; 48:109–14.

4. Setchell K. D. R. Discovery and potential clinical importance of mammalian lignams. *Flaxseed and Human Nutrition.* Cunnane S. C., Thompson L. U., eds. Champaign, IL; AOCS Press; 1995. pp. 82–98.

5. Cunnane S. Nutritional attributes of dietary flaxseed oil—Reply. *Am. J. Clin. Nutr.* 1995; 62:841–2.

6. Jenab M., Thompson L. U. The influence of flaxseed and lignans on colon carcinogenesis and beta-glucuronidase activity. *Carcinogenesis* 1996; 17:1343–8.

7. Mantzioris E., James M. J., Gibson R. A., Cleland L. G. Nutritional attributes of dietary flaxseed oil. *Am. J. Clin. Nutr.* 1995; 62:841.

8. Freeman T. P. Structure of flaxseed. *Flaxseed in Human Nutrition.* Cunnane S. C., Thompson L. U., eds. Toronto, Canada: AOCS Press.; 1995. pp. 11–21.

9. Erskine A. J., Jones J. K. N. The structure of linseed mucilage. Part I. *Can. J. Chem.* 1957; 35:1174–82.

10. Muralikrishna G., Salimath P. V., Tharanathan R. N. Structural features of an arabinoxylan and a rhamnogalacturonan derived from linseed mucilage. *Carbohydr. Res.* 1987; 161:265–71.

11. Susheelamma N. S. Isolation and properties of flaxseed mucilage. *J. Food Sci.* 1987; 24:103–6.

12. Wannerberger K., Nylander T., Nyman M. Rheological and chemical-properties of mucilage in different varieties from linseed (*Linum usitatissimum*). *ACTA Agriculturae Scandinavica* 1991; 41:311–9.

13. Mazza G., Biliaderis C. G. Functional-properties of flax seed mucilage. *Journal of Food Science* 1989; 54:1302–5.

14. Fedeniuk R. W., Biliaderis C. G. Composition and physicochemical properties of linseed (*Linum usitatissimum* L) mucilage. *Journal of Agricultural and Food Chemistry* 1994; 42: 240–7.

15. Sanftleben J. E. Mucilaginous extracts from seeds such as flaxseed. U.S. Patent 1,841,763. 1932.

16. Cui W., Mazza G., Oomah B. D., Biliaderis C. G. Optimization of an aqueous extraction process for flaxseed gum by response-surface methodology. *Food Science and Technology—Lebensmittel-Wissenschaft & Technologie* 1994; 27:363–9.

17. Mason C. T., Hall L. A. New edible colloidal gum. *Food Industries* 1948; 20:382.

18. Tomoda G., Asami Y. Mucilage from linseed or linseed meal. Japan Patent 3359. 1950.

19. Cui W., Mazza G. Methods for dehulling of flaxseed, producing flaxseed kernels and extracting lignans and water-soluble fibre from the hulls. Canadian Patent Pending, Application No. 2167951. 1996.

20. Bhatty R. S., Cherdkiatgumchal P. Compositional analysis of laboratory-prepared and commercial samples of linseed meal and of hull isolated from flax. *JAOCS* 1990; 67:79–84.

21. Cui W., Kenaschuk E., Mazza G. Influence of genotype on chemical-composition and rheological properties of flaxseed gums. *Food Hydrocolloids* 1996; 10:221–7.

22. Cui W., Mazza G., Biliaderis C. G. Chemical-structure, molecular-size distributions, and rheological properties of flaxseed gum. *J. Agric. and Food Chem.* 1994; 42:1891–5.

23. Oomah B. D., Kenaschuk E. O., Cui W. W., Mazza G. Variation in the composition of water-soluble polysaccharides in flaxseed. *J. Agric. and Food Chem.* 1995; 43:1484–8.

24. Fedeniuk R. W. Compositional analysis and physical properties of water-soluble polysaccharides from linseed. M.Sc. Thesis, University of Manitoba, Winnipeg, Canada 1993.

25. Hunt K., Jones J. K. N. The structure of linseed mucilage. Part II. *Can. J. Chem.* 1962; 40:1266–79.

26. Cui W., Mazza G. Physicochemical characteristics of flaxseed gum. *Food Res. Intl.* 1996; 29:397–402.

27. BeMiller J. N. Quince seed, psyllium seed, flaxseed, and okra gums. *Industrial Gums*, 2nd ed. Whistler R. L., BeMiller J. N., eds. New York: Academic Press; 1973. pp. 331–7.

28. Stewart S. Effect of flaxseed gum on muffin and salad dressing quality and stability. Msc. Thesis, University of Manitoba 1997.

29. Launay B., Doubblier J. L., Cuvelier G. Flow properties of aqueous solutions and dispersions of polysaccharide. In *Functional Properties of Food Macromolecules*. Mitchell J. R., Ledward D. A., ed. London and New York: Elsevier Applied Science Publishers; 1986. pp. 1–78.

30. Wallingford L., Labuza T. P. Evaluation of the water binding properties of food hydrocolloids by physicochemical methods and in a low fat meat emulsion. *J. Food Sci.* 1983; 48:1–5.

31. Gaonkar A. G. Surface and interfacial activities and emulsion characteristics of some food hydrocolloids. *Food Hydrocolloids* 1991; 5:328–37.

32. Garden-Robinson J. Flaxseed gum: extraction, composition, and selected applications. *Proceedings of the 55th Flax Institute of the United States* 1994; pp. 154–65.

33. Aliev R. K. New galenical preparations from flax seeds. *Am. J. Pharm.* 1944; 118:439.

34. Boichinov A., Akhtardzhiev K., Kolev D. A mucous substance (polyuronide) for medicinal and technical purposes obtained from linseed groats. *Chem. Abstr.* 1967; 66:68861z.

35. Tufegdzic N., Tufegdzic E., Georgijevic A. Floculation and stabilization of barium sulfate suspensions by hydrophilic colloids. *Chem. Abstr.* 1965; 62:8944f.

36. Porebski J. Tablets with prolonged action prepared using vegetable mucilage. *Chem. Abstr.* 1975; 83:136845a.

37. Siehrs E. A. U.S. Patent 2267624. 1941.

38. *Drug and Cosmetic Catalog*, 12th Ed. New York; Drug and Cosmetic Industry; 1956. p. 11957.

39. Telminski M., Olszewski Z., Plonka B., Wenda L., Bednarska K. Adhesive for increasing the adhesion of a complete denture (Polish Patent PL 109,712 B1). *Chem. Abstr.* 1982; 96:91696c.

40. Minkov E., Bogdanova S. V., Penovska T. Linseed mucilage as a water-in-oil-type emulsifier. *Chem. Abstr.* 1973; 79:45747w.

41. Nikkila O. E. Egg white substitutes [Finn. Patent 34,558(1965)]. *Chem. Abstr.* 1966; 65:11250h.

42. Schormuller J., Winter H. *Nabrung* 1958; 2:83.

43. Hervey L. R. B. Fiber treatment in papermaking. U.S. Patent 3102838. 1963.

44. Attstrom R., Galntz P. O., Hakansson H., Larsson K. Saliva substitute. U.S. Patent 5,260,282. 1993.

# CHAPTER 3

# Cereal Non-Starch Polysaccharides I: $(1\rightarrow3)(1\rightarrow4)$-β-D-Glucans

## 1. INTRODUCTION

**T**HE $(1\rightarrow3)(1\rightarrow4)$-β-D-glucans are cell wall polysaccharides of cereal endosperm and aleurone cells. The content of β-D-glucans in cereals follows the order of barley, 3–11%, oat, 3.2–6.8%, rye 1–2% and wheat, <1%. β-D-glucan is evenly distributed across the grains of barley, oat and rye except for some low β-D-glucan content oat varieties, in which β-D-glucan is more concentrated in the subaleurone layer [1,2]. In contrast, wheat β-D-glucan is primarily in the subaleurone layer with little in the endosperm cell walls. In barley, the aleurone cell walls consist of 67% arabinoxylans and 26% β-D-glucan, whereas the endosperm cell wall material contains about 20% arabinoxylans and 70% β-D-glucans [2].

Barley β-D-glucan has been well documented because of its significance in malting and brewing processes [3,4]. Improper malting or mashing will lead to high levels of β-D-glucan, and therefore, high viscosity; this may lower the extract yield and affect wort run-off. During the storage period, β-D-glucan can form gels at low temperatures that may block the filtration media in the final filtration of the beer. In the feed industry, high viscosity will reduce feed performance in chickens [5–7].

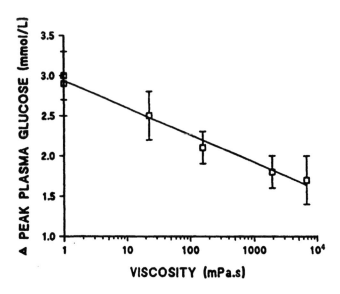

**Figure 3.1** Relationship between mean peak plasma glucose increment (peak plasma glucose ± SEM) and log η in healthy subjects consuming a 50 g oral glucose load in 500 mL water in the presence of oat gum at doses 0 (control) and 1.8, 3.6, 7.2 and 14.5 g [14].

Increasing interests in oat β-D-glucan during the last two decades are largely due to its beneficial physiological effect on human health. In 1997, the Food and Drug Administration (FDA) of the U.S. allowed a health claim for an association between consumption of diets high in oat β-D-glucan products, such as oatmeal, rolled oat and oat bran, and reduced risk of coronary heart disease [8]. In the claim, β-D-glucan was accepted as the main active ingredient in these oat products. Of the numerous physiological studies in the literature, an outstanding contribution was made by Wood and coworkers [9–11]. They found that the ability of oat β-D-glucan to reduce serum blood cholesterol and glycemic index was apparently correlated to the viscosity produced by the β-D-glucans as shown in Figure 3.1. The viscosity of β-D-glucan solution is determined by its primary structure, molecular weight and distribution and concentration of solutions.

Studies on rye and wheat β-D-glucans are limited because of their low contents in the grains. However, our laboratory developed some interest in wheat β-D-glucan due to a preprocessing technology that enriched β-D-glucan content from 0.5% in whole wheat to 2.6–3.0% in one of the bran fractions [12–14].

Generally, cereal β-D-glucans share a common structural feature: there are consecutive blocks of (1→4)-linked β-D-glucose (mostly two or three, and sometimes up to 14) that are separated by a single (1→3)-linked β-D-glucose, as shown in Figure 3.2 [15]. However, the fine structure and molecular weight of

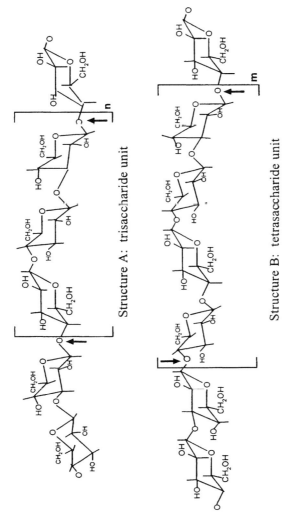

Structure A: trisaccharide unit

Structure B: tetrasaccharide unit

**Figure 3.2** Main structural features of cereal β-D-glucans. $M \approx 25$–$35\%$, $n \approx 65$–$75\%$. Reprinted from Reference [15].

β-D-glucans from different cereal origins showed clear patterns of difference. As a result, the physical properties of these polysaccharides, such as rheological behavior and gelling properties, are different. These variations in structural features and rheological behavior might have significant impact on their physiological functionality and application in the food and cosmetic industries. The goal of this chapter is to review the most recent advances in processing, molecular structure, solution and functional properties of cereal β-D-glucans and their potential uses as stabilizers and moisturizers in food and non-food applications.

## 2. EXTRACTION, PURIFICATION AND DETERMINATION

### 2.1. SOLUBILITY AND EXTRACTABILITY

In addition to structural features and molecular weight/molecular weight distribution, solubility of β-D-glucan is another important determinant that not only affects the extractability from raw material but also affects the functionalities in food systems and in the human body. However, the solubility of β-D-glucan is dependent on fine structure and molecular weight. Most β-D-glucans contain approximately 30% (1→3)- and 70% (1→4)-linkages that are organized into blocks of two or three consecutive (1→4)-linked residues separated by (1→3)-linked residues [15] (Figure 3.2). The higher solubility of mixed-linked β-D-glucans compared to cellulose is attributed to the presence of irregular (1→3)-β-bonds.

The solubility of β-D-glucan in water could be defined as the relative maximum amount of β-D-glucan that could be introduced to "solution." Since it is difficult to obtain a true solution of macromolecules like β-D-glucans, solubility is also referred to as extractability under certain specified conditions during sample preparation, such as solvent, temperature, time, liquid:solids ratio, etc. Because of the complexity and the many variables involved, it is not practical to compare data from different studies regarding the solubility and/or extractability of β-glucan from cereals [16–20].

Solubilization of β-D-glucan in aqueous solvents involves heating at different temperatures for certain time periods. Extraction of cereal β-glucan with mild reagents and conditions is incomplete. McCleary [18] reported that ~90% of the total barley β-glucan was extracted in successive treatments with water at 40, 65 and 95°C. Whereas, Wood et al. [17] found only ~45% of barley and ~70% of oat β-glucans extracted by carbonate (pH 10) at 60°C. Beer et al. found that ~65% of the total β-glucan in oats and waxy barley and ~50% in non-waxy barley can be extracted by hot water (90°C) containing a thermostable, β-glucanase-free, α-amylase [21]. Additional hot water extractions and a final DMSO extraction increased the total β-glucan solubilized to ≈80%. A total extraction of β-D-glucans is achievable using sodium hydroxide. Alkali

(NaOH) and acids (dilute perchloric or sulphuric acid) have been used for complete extraction, which is necessary for analysis; however, both solvents lead to depolymerization of β-D-glucans [22–24].

Varum and Smidsrod [25] observed that for oat aleurone, at pH 9.2, higher temperatures gave extracts with higher intrinsic viscosities, which may indicate that higher molecular weight β-D-glucans have been extracted. However, a controversial result was reported for β-D-glucan from barley endosperm [26]. In this study, the β-D-glucan extracted with water at 65°C was less soluble than β-D-glucans extracted at 40°C, although the molecular weight of β-D-glucan extracted at 65°C was lower compared to that at 40°C. The freeze-dried 65°C water-soluble β-D-glucan proved difficult to dissolve in aqueous buffer, requiring several hours of constant stirring at 90°C before dissolution was completed. In contrast, the 40°C water-soluble β-D-glucan dissolved after 1–2 hr at 70°C [26,27]. The 65°C water-soluble fraction also showed a greater tendency to precipitate from solution when the temperature was lowered. Freeze-dried wheat β-D-glucan, which has a more regular structure compared to barley β-D-glucan, exhibited extreme insolubility in water, even heated at 100°C for several hours. However, wheat β-D-glucans prepared by the solvent exchange method [12,28] readily dissolved in water upon heating at 90°C for 2 hr. It is demonstrated that the structure regularity of cereal β-D-glucans follows the order of wheat ≫ barley (65°C) > barley (40°C) > rye > oat. The solubility of cereal β-D-glucans follows the reversed order of the structure regularity.

Under mild extraction conditions, only 30–70% of the total β-D-glucan can be extracted [29]. Under vigorous conditions, it appeared that a total extraction could be achieved, however, the β-D-glucan might be depolymerized. A thorough study was carried out recently by Beer et al. to examine the effects of extracting conditions/solvents on the extractability and molecular weight of β-D-glucan from oat and barley varieties, as demonstrated in Figure 3.3. In this study, β-D-glucan was extracted consecutively from oat and barley samples by three water extractions at 90°C, each for 2 hr (heat-stable α-amylase was added to the first extraction to depolymerize starch), followed by DMSO or NaOH/NaBH$_4$ (5%/0.5%) at 22°C for 16 hr. About 60–65% of total β-D-glucan was extracted from oat and waxy barley samples in the first extraction with water and heat-stable α-amylase, while the following two water extractions gave an additional 10–15% of the total β-D-glucan. However, only 50% of the total β-D-glucan was extracted from the non-waxy barley varieties, as shown in Table 3.1. DMSO gave another 6–8% of the total β-D-glucan, while NaOH/NaBH$_4$ solutions produced 24–35% of total β-D-glucan, which almost achieved total extraction (86–105%). The average peak molecular weights of these samples are presented in Table 3.2. No changes in peak MW in the first two water consecutive extracts E1, E2 and the DMSO extract were detected. However, significantly lower MW was found in the E3 extracts, possibly due to degradation of the polymer during extraction, or a low MW fraction of

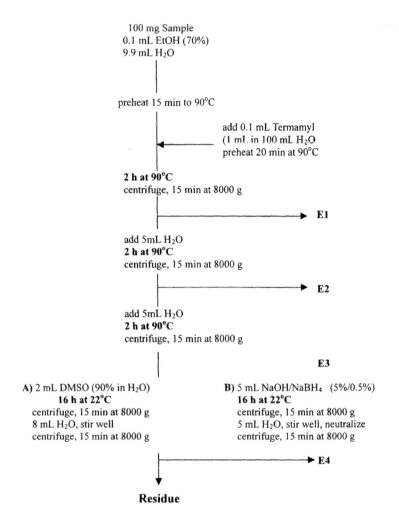

**Figure 3.3** Sequential extraction procedure of β-D-glucans from oat and barley using water, NaOH/NaBH₄ and dimethyl sulfoxide (DMSO). Reprinted from Reference [21].

β-D-glucan was extracted. 1 M NaOH solution (with NaBH₄) (22°C, 16 hr) depolymerized β-D-glucan to a substantial extent, although it gives apparent total extraction [21].

In the framework of evaluating the nutritional and physiological aspects of cereal β-glucans, an *in vitro* extraction system, which mimics human digestion, was developed to evaluate the effects of processing and cooking on the solubility of β-glucan [29]. Bran samples were compared with samples that went through a baking process. The extraction was carried out at 37°C and pH 6.9 for ∼2 hr.

TABLE 3.1. Solubility/Extractability of β-D-Glucan from Oat and Barley Varieties in Different Solvents.

| Sample | β-Glucan (g/100 g) | Consecutive Extracts (% of Total Amount) | | | | | | |
| | | H₂O | | | DMSO E4 | NaOH E4 | E4 DMSO Total | E4 NaOH Total |
| | | E1 | E2 | E3 | | | | |
| --- | --- | --- | --- | --- | --- | --- | --- | --- |
| Oat | | | | | | | | |
| AC Lotta | 5.89 | 58.5 | 10.04 | 3.03 | 7.17 | 21.7 | 78.83 | 91.9 |
| Capitol | 6.23 | 61.66 | 8.93 | 2.16 | 6.07 | 13.85 | 80.15 | 85.7 |
| Rigodon | 5.02 | 66.2 | 9.90 | 2.75 | 6.57 | 14.25 | 83 | 93.65 |
| AC Steward | 6.06 | 57.54 | 11.09 | 3.58 | 6.91 | … | 76.05 | … |
| Newman | 5.38 | 63.43 | 9.31 | 2.96 | 7.88 | … | 78.68 | … |
| Marion | 6.08 | 65.81 | 9.06 | 2.85 | 7.22 | … | 81.3 | … |
| Mean | 5.77 | 62.18 | 9.72 | 2.89 | 6.96 | 16.6 | 79.67 | 90.42 |
| Barley (non-waxy) | | | | | | | | |
| Compana | 5.79 | 50.03 | 7.37 | 2.97 | 7.44 | 33.67 | 62.35 | 92.05 |
| Nupana | 5.38 | 53.77 | 8.09 | 3.19 | 8.29 | 33.32 | 65.7 | 97.9 |
| Shopana | 4.71 | 49.13 | 7.24 | 2.7 | 7.71 | 33.75 | 59.3 | 92.6 |
| Shonupana | 5.71 | 50.36 | 7.36 | 3.04 | 6.00 | 36.84 | 61.25 | 97.1 |
| Mean | 5.4 | 50.83 | 7.53 | 2.98 | 7.36 | 34.39 | 62.15 | 94.91 |
| Barley (waxy) | | | | | | | | |
| Wapana | 6.15 | 64.9 | 11.16 | 3.4 | 6.66 | 21.41 | 78.65 | 101.75 |
| Wanupana | 6.69 | 65.34 | 8.44 | 2.82 | 5.27 | 23.8 | 71.1 | 105.85 |
| Washopana | 6.23 | 59.96 | 6.98 | 3.06 | 6.03 | 22.71 | 72.6 | 90.1 |
| Washonupana | 8.07 | 59.65 | 7.59 | 4.07 | 6.92 | 23.86 | 72.8 | 93.65 |
| Mean | 6.79 | 62.46 | 8.54 | 3.34 | 6.21 | 22.95 | 73.79 | 97.84 |

Reprinted from Reference [21].

TABLE 3.2. **Comparison of Peak Molecular Weights (Mp $\times$ 10$^{-3}$)
of Consecutive Extracts from Oat and Barley.**

| | $H_2O$ | | | | |
|---|---|---|---|---|---|
| **Sample** | **E1** | **E2** | **E3** | **E4 DMSO** | **E4 NaOH** |
| Oat | | | | | |
| AC Lotta | 2485 | 2846 | 1029 | 2694 | 1010 |
| Capitol | 2510 | 2123 | 647 | 2672 | 1067 |
| Rigodon | 2287 | 2226 | 333 | 1886 | 1227 |
| AC Steward | 2469 | 2581 | 1398 | 1908 | . . . |
| Newman | 2269 | 1947 | 645 | 1671 | . . . |
| Marion | 2094 | 2032 | 528 | 1428 | . . . |
| Mean | 2346 | 2329 | 788 | 1918 | 1102 |
| Barley (non-waxy) | | | | | |
| Compana | 1685 | 1425 | 723 | 1919 | 1244 |
| Nupana | 1647 | 1465 | 671 | 1919 | 1157 |
| Shopana | 1686 | 1673 | 494 | 1937 | 1097 |
| Shonupana | 1643 | 1722 | 428 | 1931 | 1107 |
| Mean | 1665 | 1571 | 579 | 1926 | 1151 |
| Barley (waxy) | | | | | |
| Wapana | 1315 | 1422 | 829 | 1918 | 1151 |
| Wanupana | 1317 | 1680 | 601 | 1968 | 1034 |
| Washopana | 1512 | 1794 | 911 | 2310 | 1088 |
| Washonupana | 1398 | 2115 | 1230 | 2496 | 1124 |
| Mean | 1386 | 1754 | 893 | 2181 | 1100 |

Reprinted from Reference [29].

Only 13–29% of the total β-D-glucans were extracted from the original bran sample. The extractable β-D-glucans from cooked samples increased to 30–85% of the total β-D-glucans, as shown in Table 3.3; the increase in the extracting rate was thought to be caused by the degradation of β-D-glucans [29].

There is as yet no fully satisfactory explanation for the resistance to the solubilization of β-glucan, although it is evident that MW may have a role [14,19]. β-D-Glucans in thicker cell walls, such as in the subaleurone endosperm of many cultivars, show a greater resistance to extraction [30]. However, wheat β-glucan, despite the thin endosperm cell walls, is extremely resistant to extraction [31]. Evidence has shown that the chemically cross-linked ferulic acid-arabinoxylan complex in the aleurone cell walls is the primary reason for the resistance to solubilization: wheat bran arabinoxylans and β-glucans are not extractable in hot water but are soluble in alkali solvent at room temperature [12,28]. In the latter case, alkali solvent breaks the ester bond of the cross-links between arabinoxylans and ferulic acid and, therefore, solubilizes the non-starch polysaccharides in aqueous solution. There is no evidence that β-glucan is chemically linked

TABLE 3.3. **Comparison of Extraction Rates and Molecular Weights of β-D-Glucans from Oat Bran Samples Using Hot Water and Physiological Conditions.**

| Sample | β-Glucan (g/100 g) | β-Glucan Extracted (% of Total) | | Molecular Weight ($\times 10^{-3}$) | |
|---|---|---|---|---|---|
| | | Hot Water | Physiological | Hot Water | Physiological |
| Bran A | 13.4 | 51.3 | 12.9 | 1400 | 1100 |
| Bran B | 8.9 | 56.7 | 25.1 | 1600 | 1800 |
| Bran C | 7.6 | 64.1 | 28.7 | 1800 | 1900 |
| Bran C[a] | nd[b] | 30.2 | — | 1800 | — |
| Rolled oats[a] | 4.2 | 69.5 | 33.4 | 1500 | 1500 |

[a]Samples cooked before extraction.
[b]Not determined.
Reprinted from Reference [29].

to ferulic acid or to arabinoxylans. Therefore, the resistance of wheat β-glucan to solubilization is possibly due to its physical entrapment in the matrix of the cross-linked ferulic acid-arabinoxylan complex. Because alkaline conditions permit total extraction of β-D-glucans from oat and barley [29], it is likely that the insoluble portion of β-D-glucan in these cereals was also entrapped in the cross-links of phenolic components and arabinoxylans; however, evidence is needed to prove this hypothesis.

In practical applications, β-D-glucans are usually co-extracted with arabinoxylans. The extraction process generally has three steps: (a) treatment with hot aqueous-alcohol solvent to deactivate any endogenous enzymes present and to remove impurities such as free sugars, small proteins and some non-polar compounds; (b) incubation with saliva α-amylase or thermally stable α-amylase to depolymerize starch—this step simultaneously extracts the non-starch polysaccharides (pentosans and β-D-glucans); and (c) purification of the extracted β-D-glucans by adding $(NH_4)_2SO_4$ or using enzymes to remove or depolymerize the arabinoxylans. In the following three sections, extraction and purification of β-glucans from three cereals are described in detail.

## 2.2. EXTRACTION FROM OAT

An extraction and fractionation procedure described by Westerlund et al. is shown in Figure 3.4. This procedure gives β-D-glucan and arabinoxylans from milled oat fractions. First, a mixture of hot isopropanol and petroleum ether (2:3, v/v) was used to inactivate endogenous β-glucanase and remove lipophilic components. An extraction with 90% aqueous ethanol further removed polar

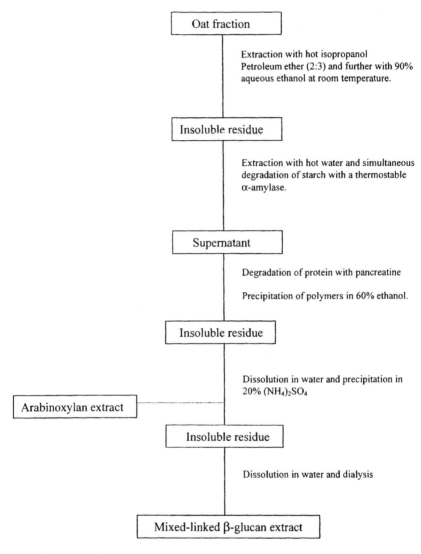

**Figure 3.4** Scheme for the isolation of arabinoxylan and mixed-linked β-glucan extracts from different oat fractions. Reprinted from Reference [32] with permission Elsevier Science.

substances such as low molecular weight sugars (oligomers and monomers) and some proteins. The starch was degraded by treatment with a thermally stable α-amylase (Termamyl 120 L, Novo Nordisk A/A, Copenhagen, Denmark). The removal of lipophilic substances was considered necessary to optimize the yield and purity of the isolated arabinoxylan and β-D-glucan [32]. The supernatant from the hot water extraction was treated with pancreatine to decrease protein content, then the non-starch polysaccharide mixture was precipitated in 60% aqueous ethanol, and the oligosaccharides from the starch degradation remained in the solution. The aqueous ethanol precipitate was further fractionated by dissolution in water, and the β-D-glucans were precipitated with 20% $(NH_4)_2SO_4$. The arabinoxylans and β-D-glucan were recovered from the supernatant and precipitated, respectively, dialyzed against distilled water and freeze-dried. The authors noted that the precipitation of the extract in 60% aqueous ethanol was necessary for removing the oligosaccharides from degraded starch. If 20% $(NH_4)_2SO_4$ was added directly to the non-starch polysaccharide mixture, the incomplete precipitation of the β-D-glucan resulted in contamination of the arabinoxylan fraction [32].

A procedure developed by Wood and coworkers [17,30,33–35] involved mild alkaline extraction (sodium carbonate buffer pH 10, I = 0.2) at 45°C for 30 min and removal of co-extracted protein by adjusting the pH to 4.5 followed by centrifugation. This method produced a non-starch polysaccharide mixture of β-D-glucans and arabinoxylans with 5% protein contaminant. The non-starch polysaccharide mixture is further fractionated into arabinoxylans and β-D-glucan by the $(NH_4)_2SO_4$ precipitation method developed by Preece and Hobkirk in 1953 [33,36]. Based on the same principles, Wood and coworkers [34] also reported a pilot plant procedure for preparation of oat gum (78% β-D-glucan). This procedure includes two steps. The first step is to prepare a β-D-glucan-rich bran fraction using a roller mill or pin mill, followed by air classification/screening. The prepared bran is refluxed in ethanol (75%, v/v) to deactivate endogenous enzymes. Usually, 55 kg of bran is mixed with 500 L of aqueous ethanol (75%, v/v) in a 560-L reactor vessel prepared for reflux circulation at ~80°C and stirred for 4 hr. The cooled (50°C) contents are then pumped to a plate and frame filter press and purged of free ethanol with $N_2$. The wet bran is removed and reslurried in 100% ethanol and stirred for 15 min, then transferred to a fresh filter press and purged with $N_2$. The wet bran is allowed to desolventize for three days in a stainless steel, forced-ambient-air desolventizer.

The extraction process of oat gum from the bran obtained is demonstrated in Figure 3.5. Deactivated bran (25 kg) is blended with water (500 L) at 33–35°C using a model DAS06 comminuting mill (Fitzpatric Co., Elmhurst, IL) equipped with a 1.6 mm screen. The pH is adjusted to 10 by additing 20% sodium carbonate. The mixture is stirred in a holding tank for 30 min. Liquid extract is then separated from the residue using a 15 × 30 cm decanting centrifuge (Bird Machine, S. Walpole, MA) operating at 5000 rpm and a flow rate of 4.6 L/min. Solids are bagged and held at −20°C prior to further extraction. Frozen bran is ground in a meat grinder fitted with a 4.5-mm die (Hobart Mfg., Troy, OH) prior to extraction with 500 L of pH 10 sodium carbonate at ~45°C. The liquid extract is cooled by pumping through a CV-5 shell and tube chiller at 18°C

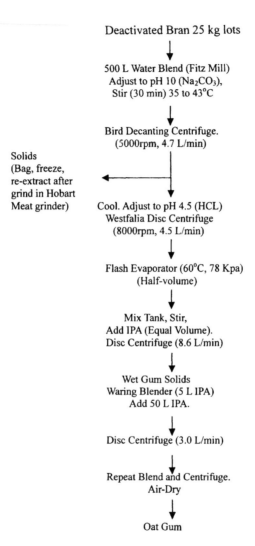

Deactivated Bran 25 kg lots

500 L Water Blend (Fitz Mill)
Adjust to pH 10 (Na₂CO₃),
Stir (30 min) 35 to 43°C

Bird Decanting Centrifuge.
(5000rpm, 4.7 L/min)

Solids
(Bag, freeze,
re-extract after
grind in Hobart
Meat grinder)

Cool. Adjust to pH 4.5 (HCL)
Westfalia Disc Centrifuge
(8000rpm, 4.5 L/min)

Flash Evaporator (60°C, 78 Kpa)
(Half-volume)

Mix Tank, Stir,
Add IPA (Equal Volume).
Disc Centrifuge (8.6 L/min)

Wet Gum Solids
Waring Blender (5 L IPA)
Add 50 L IPA.

Disc Centrifuge (3.0 L/min)

Repeat Blend and Centrifuge.
Air-Dry

Oat Gum

**Figure 3.5** Flow chart summarizing pilot plant production of oat gum. Reprinted from Reference [34] with permission from *Cereal Chemistry*. © 1989. The American Association of Cereal Chemists, Inc.

(Chester Jensen, Chester, PA) to a holding tank where the pH is adjusted to 4.5 with 20% (v/v) hydrochloric acid. The resulting mixture is centrifuged in a Westfalia disk centrifuge (SA-7-01-576, Centrico, Northvale, NJ) at 8400 rpm and 4.5 L/min with the desludge cycle set at 10 min intervals. The supernatant is pumped to a flash evaporator Model R56, APV-Crepaco, Tonawanda, NY) and reduced to 40–50% of original volume at ~4.2 L/min with a heat exchange temperature of 60°C and a system pressure of 23 kPa. The liquid is then cooled and pumped to a mixing tank, and an equal volume of 100% propan-2-ol or ethanol is added with vigorous stirring. The precipitated gum is collected under nitrogen in the disk centrifuge at a feed rate of 8.6 L/min with desludge intervals of 20 min. The gum solids are blended with 100% propan-2-ol or ethanol in a 5-L Waring Blendor and made up to 50 L with alcohol. The mixture is separated at 3 L/min in an explosion-proof decanting centrifuge (Bird, 15 × 30 cm) at a speed of 5000 rpm. The gum cake is, again, blended, suspended in alcohol, centrifuged and then air desolventized for two to three days and ground in a Jacobsen hammer mill (Model 66B, Kipp-Kelly) equipped with a 250-μm screen. The air desolventization can be speeded up by gentle heating. Using this procedure, Wood and coworkers extracted 18.6 kg of oat gum from 2000 kg of oat bran [34]. This material has 78% β-D-glucan with about 8% starch contaminates. Further purification by the ammonium sulphate precipitation method increased the β-D-glucan content to 90–95% [34].

## 2.3. EXTRACTION FROM BARLEY

Commercial barley β-D-glucan is prepared according to a method developed by McCleary [18]. The basic principles are the same as in the extraction of oat β-D-glucan, which involves deactivation of endogenous enzymes, removal of starch and extraction and purification of β-D-glucan. Modification of McCleary's method was recently reported by Bohm and Kulicke [37], as described below.

Barley (150 g) is ground on a 0.5 mm sieve then refluxed with 80% (v/v) ethanol (750 mL) for 30 min. The flour is then washed with 96% (v/v) ethanol (750 mL) and dried at 80°C. A suspension of the flour in phosphate buffer (0.01 M, pH 6.0) is slowly heated to 95°C. When the starch gelatinization starts (produces high viscosity at about 70°C), heat-stable β-amylase is added and kept at 95°C for 1 hr. The starch is degraded into small dextrins after 15 min, as indicated by iodine staining. The residue is removed by centrifugation and washed twice with water (250 mL). The washes are added to the extract, and the volume is reduced to 500 mL by diafiltration. Ammonium sulphate is added to the concentrated extract to give a final concentration of 30% (w/v), and the sample is kept overnight at 4°C. The precipitate is centrifuged and washed with 20% (v/v) and 96% (v/v) ethanol (100 mL), respectively, and resuspended in deionized water with a kitchen homogenizer, then dissolved (1000 mL) at 80°C. The solution is filtered by celite (Sigma, acid wash) to remove hazes. An equal volume of 96% (v/v) ethanol is added to the filtrate to precipitate β-D-glucans. The precipitate is washed again with 96% (v/v) ethanol and dried in vacuum. A yield of 2.3% high molecular weight β-D-glucan is obtained by this method (the β-D-glucan content was 5.1% in the grain).

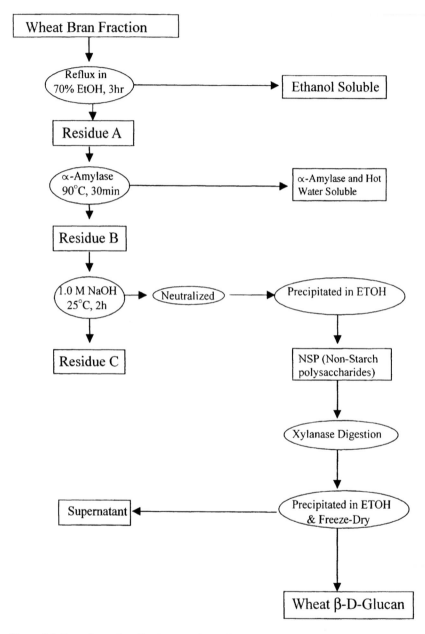

**Figure 3.6** Extraction and purification procedures of wheat β-D-glucan from preprocessed wheat bran.

## 2.4. EXTRACTION FROM WHEAT BRAN FRACTION

Because there is only 0.5–1% β-D-glucan in wheat, it might not be feasible to choose the whole grain as a source of β-D-glucan. Nevertheless, a commercial wheat preprocess technology produced a bran fraction (fraction B) that contained 2.5–3% of β-D-glucan [28]. An extraction and purification procedure was developed by Cui and coworkers, as described below (Figure 3.6): the bran fraction is first treated with 70% ethanol at 70°C for 3 hr to deactivate the endogeneous enzymes and remove some soluble materials, including free sugars, amino acids and some phenolics. The residue from ethanol extraction (residue A) is then digested by thermally stable α-amylase to eliminate starch that might contaminate the nonstarch polysaccharides (NSP) in a later stage. Arabinose, xylose β-D-glucan were not detected in the α-amylase treated liquid phase. This result suggests that there were no β-glucans soluble in hot water. Therefore, the starch in Residue A can be removed without losing the NSP. The NSP, a mixture of arabinoxylans and β-D-glucans, is then extracted from starch-free wheat bran fraction B with 1.0 M NaOH at 25°C for 2 hr (solid to liquid ratio = 1:10) [12,28]. Alkaline extracts are neutralized with 2 M HCl and centrifuged at 5000 g at 25°C for 20 min. The supernatant is adjusted to pH 4.75 with 0.25 M sodium acetate buffer and treated with xylanase (~100 unit/100 mL of extract) at 50°C under constant stirring for 2 hr. The enzyme is deactivated by heating at 80°C for 30 min, and the solution is centrifuged at 5000 g at 25°C for 20 min. The supernatant solution is made to 50% ethanol (final concentration), and the precipitate is recovered by centrifugation. The resulting precipitate is washed with 50% ethanol, then suspended in 100% 2-propanol and kept at 4°C overnight. After removal of solvent, the precipitate is dried with gentle warming [34]. The gradual ammonium sulphate precipitation technique was also used to purify β-D-glucan; however, due to the low level of β-D-glucan in the NSP, the purification procedure proved to be tedious and time consuming.

## 3. STRUCTURE OF β-D-GLUCANS

## 3.1. STRUCTURAL FEATURES

β-D-Glucan is a linear, unbranched polysaccharide containing a single type of monosaccharide, i.e., β-D-glucose. The β-D-glucose has two types of linkage, one is (1→4)-*O*-linked β-D-glucopyranosyl unit (~70–72%) and the other is (1→3)-*O*-linked β-D-glucopyranosyl (~28–30%) unit. The structure features the presence of consecutive (1→4)-linked β-D-glucose (mostly 2 or 3, and sometimes up to 14 blocks) with blocks that are separated by single

**TABLE 3.4.** Comparison of Structural Features of β-Glucans from Three Cereals after Lichenase Hydrolysis [15].

| β-D-Glucan Source | Peak Area Percent % | | Total % | | Molar Ratio |
|---|---|---|---|---|---|
| | Tri[a] | Tetra[a] | Tri + Tetra[b] | Penta → Nona[b] | Tri/tetra[c] |
| Wheat (preprocessed) | 72.3 | 21.0 | 93.3 | 6.7 | 4.5 |
| Wheat bran (AACC, soft white) | 70.6 | 22.9 | 93.5 | 6.5 | 4.2 |
| Wheat bran (AACC, red hard) | 72.0 | 21.0 | 93.0 | 7.0 | 4.5 |
| Barley | 63.7 | 28.5 | 92.2 | 7.8 | 3.0 |
| Oat | 58.3 | 33.5 | 91.9 | 8.1 | 2.3 |

[a]HPLC peak area percentage of tri- and tetrasaccharides after lichenase hydrolysis.
[b]Total percentage of tri- and tetrasaccharides and penta- to nonasaccharides, respectively.
[c]The molar ratio of trisaccharide over tetrasaccharide from lichenase hydrolysis.

$(1{\rightarrow}3)$-linkage, as shown in Figure 3.2. The $\beta$-$(1{\rightarrow}3)$-linked cellobiosyl unit (trisaccharide, unit A) is the major building block (58–72%), followed by $\beta$-$(1{\rightarrow}3)$-linked cellotriosyl unit (tetrasaccharide, unit B, 20–34%). The $(1{\rightarrow}4)$ linkages occuring in a group of greater than three are also present, but in an extremely lower amount as summarized in Table 3.4. The structural features of $\beta$-glucans from different origins are very similar as studied by $^{13}$C-NMR; the spectra of $\beta$-D-glucan isolated from three cereals appear to be almost identical (Figure 3.7) [15]. However, significant differences are revealed by their oligosaccharide composition that is obtained by specific enzyme hydrolysis (lichenase, EC 3.2.1.73) and quantified by HPLC analysis, as detailed in section 3.2.

## 3.2. OLIGOSACCHARIDE COMPOSITION—LICHENASE HYDROLYSIS

Lichenase is a $(1{\rightarrow}3)(1{\rightarrow}4)$-$\beta$-D-glucan-4-glucanohydrolase (E.C. 3.2.1.73) that specifically cleaves the $(1{\rightarrow}4)$-linkage of the 3-$O$-substituted glucose unit in $\beta$-D-glucans. Samples containing $\beta$-D-glucan, such as whole grain, flours, brans, isolated $\beta$-D-glucans or food product containing D-glucans, can be dissolved in 0.05 M phosphate buffer (pH 6.9) then digested by lichenase. The amount of enzyme is adjusted according to the $\beta$-D-glucan content and sample size. The lichenase digest can be analyzed by high-performance anion exchange chromatography (e.g., a Dionex system with gold electrodes equipped with a PAD). Aliquots of the digest are diluted with water up to 20-fold to analyze the

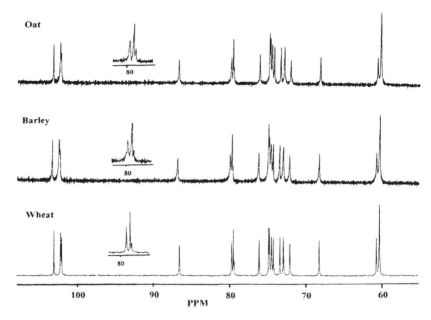

**Figure 3.7** <sup>13</sup>C-NMR spectra of three cereal β-D-glucans (4% w/v) in deuterated dimethylsulfoxide. The experiment was carried out on a Bruker WM500 NMR spectrometer at 90°C [15].

relative amount of trisaccharide and tetrasaccharide. A higher concentration is required to analyze oligosaccharide of DPs greater than 4 due to their low presence in β-D-glucan. Typical chromatographic profiles of the oligosaccharides are shown in Figure 3.8 [14].

The relative amounts of oligosaccharides produced by lichenase treatment constitute a fingerprint of the structure of a β-D-glucan. The total of tri- and tetrasaccharide of cereal β-D-glucan is similar (92–93%) among cereal β-D-glucans, however, the amount of trisaccharide follows the order of wheat (~72%), barley and rye (~66%) and oat (~58%). This is also reflected by the ratio of tri- to tetrasaccharides that follows the order of wheat (4.6), barley (3.3), rye (2.7) and oat (2.2). The ratio of tri- to tetrasaccharides can be used as a parameter that defines the fingerprint of the structure of cereal β-D-glucan—it will be used frequently in this chapter.

In the lichenase digested product, there is a precipitate observed. Because the enzyme is specific for β-(1→4)-linked 3-substituted glucopyranosyl units, these high DP oligosaccharides may contain more than nine β-(1→4)-linked glucopyranosyl units that are terminated at the reducing end by a β-(1→3)-linked glucose (i.e., cellodextrins terminated at the reducing end by a 3-linked unit) [8]. These structures were confirmed by methylation analysis and <sup>13</sup>C-NMR spectroscopy [14,25]. The high DP oligosaccharides arise from

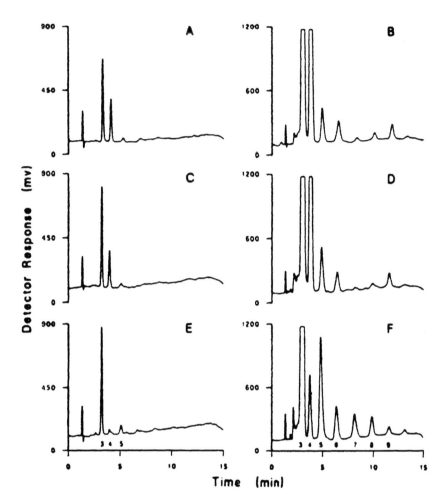

**Figure 3.8** High-performance anion-exchange chromatograms with pulsed amperometric detection of lichenase-treated oat (A, B) and barley (C, D) and β-glucan and lichenan (E, F). B, D and F have increased concentrations in order to analyze oligosaccharides of DP 5–9. Numbers 3–9 indicate oligosaccharide DP. Reprinted from Reference [14].

cellulose-like portions of β-D-glucan that might contribute to the stiffness of the molecules (and, hence, rheological behavior) in solution [25].

There were some controversial reports about the minor component of cereal β-D-glucans. There was a suggestion of consecutive β-$(1\rightarrow3)$ linkages in cereal β-glucan [38], however, it has not been confirmed. If the consecutive β-$(1\rightarrow3)$ linkages did present, it would be in minor amounts (<1%). Protein or peptide linkage has also been suggested [38], and MW and viscosity loss on incubation with a β-glucanase-free trypsin has been reported [39]. In our laboratory, it was observed that high molecular weight β-D-glucans from oat and barley were degradable by β-D-xylanase, but wheat β-D-glucan was not degradable by the same enzyme.

## 3.3. $^1$H, $^{13}$C AND TWO-DIMENSIONAL NMR SPECTROSCOPY

To further understand the structural properties of cereal β-D-glucans, Cui and coworkers carried out a detailed NMR analysis on wheat β-D-glucan. High-resolution $^1$H and $^{13}$C NMR spectra were recorded in dimethyl sulfoxide (DMSO-d6) at 500.13 and 125.78 MHz, respectively, on a Bruker AM500 NMR spectrometer operating at 90°C. Chemical shifts were referenced to DMSO-d6 at 2.49 ppm for $^1$H and 39.5 ppm for $^{13}$C and are reported relative to TMS. Homonuclear $^1$H/$^1$H correlation spectroscopy (COSY, TOCSY) and heteronuclear $^{13}$C/$^1$H correlation experiments (HETCOR, HMBC) were recorded in order to elucidate the complete assignment of the spectra.

The $^{13}$C NMR spectrum of wheat β-D-glucan at 90°C appeared to be similar to that of oat and barley, as demonstrated in Figure 3.7. The $^1$H spectrum, however, was well resolved compared to that of previous studies of such glucans [40]. It showed three distinct doublets for the anomeric protons of β-D-glucan residues in three distinctly different environments (Figure 3.9). The $^{13}$C-$^1$H heteronuclear correlation (Figure 3.9) allowed a one-to-one match of the $^{13}$C and $^1$H resonances. The $^1$H/$^1$H correlation (COSY) spectrum (Figure 3.10) [41] assigned the chemical shifts of the three H-2s from coupling with their respective anomerics, as well as the three H-3s, which were derived from their corresponding H-2s. Similarly, the H-6 AB systems and respective H-5s were assigned, but clear assignment of the H-4s was unobtainable due to a high degree of overlap between the H-3, 4 and 5 protons. With the knowledge of the direct correlations, a total $^1$H/$^1$H correlation spectrum (TOCSY) provided all of the intraresidue assignments. The subspectra from the two-dimensional TOCSY spectrum corresponding to each of the anomeric resonances are shown in Figure 3.11 with the single pulse spectrum (d) presented for comparison. Due to similar values of the $^3J_{H,H}$ for all the protons of β-D-glucose (7 to 9 Hz around the hexose ring), polarization is transferred completely from H-1 to H-6 in each residue, and the resolution of the anomerics permits the assignment of all protons within each of the three residues. Comparison of the subspectra indicated that one [(b) in Figure 3.11] is distinct from the others, which then might be assigned to the (1→3)-linked residue (**B** in Scheme 1), while the other two are attributed to the flanking (1→4)-linked residues (**A** and **C**, respectively).

Correlations of the protons from the residue tentatively assigned to the (1→3)-linked β-D-glucose residue (**B** in Scheme 1) gave carbon peaks at 101.9, 86.4, 75.9, 71.8, 68.0 and 60.5 ppm that can be assigned to C-1, C-3, C-5, C-2,

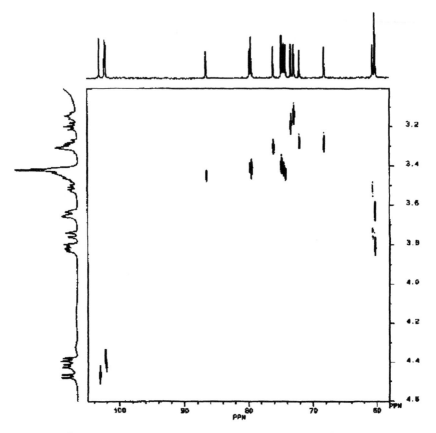

**Figure 3.9** $^{13}C$ and $^{1}H$ heteronuclear correlation NMR spectra of wheat β-D-glucan in deuterated dimethyl sulfoxide. Reprinted from Reference [41] with permission from Elsevier Science.

C-4 and C-6, respectively. These assignments confirm the (1→3) linkage. The presence of only a single resonance for C-3 of this residue demonstrates that, like oat and barley β-D-glucans, there are no consecutive (1→3)-linkages in wheat β-D-glucan [14,27]. These assignments along with the remaining correlations defining the two (1→4) linked β-D-glucan residues are summarized in Table 3.5. Overlapping resonances in the proton spectrum, however, did not permit the unambiguous assignment of the C-3 and C-5s of the (1→4)-linked residues, and the carbon and proton resonances for C-6 are coincident.

$$\overset{A}{\rightarrow}{}^{4}G_1\overset{B}{\rightarrow}[{}^{3}G_1\overset{C}{\rightarrow}{}^{4}G_1\overset{A}{\rightarrow}{}^{4}G_1\rightarrow]_m[{}^{3}G_1\overset{B}{\rightarrow}{}^{4}G_1\overset{C}{\rightarrow}{}^{4}G_1\overset{D}{\rightarrow}{}^{4}G_1\rightarrow]_n{}^{3}G_1\overset{B}{\rightarrow}{}^{4}G_1\overset{C}{\rightarrow}$$

**Scheme 1** model structure of wheat β-D-glucan

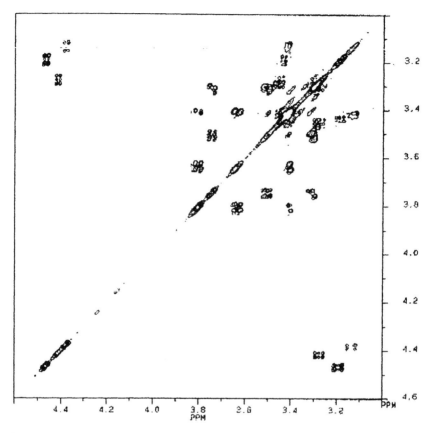

**Figure 3.10** Correlated spectroscopy (COSY) spectra of wheat β-D-glucan in deuterated dimethyl sulfoxide. Reprinted from Reference [41] with permission from Elsevier Science.

Information to resolve ambiguities in $^{13}$C assignments and to establish the sequence of the three β-D-glucose residues was obtained from analysis of the long range heteronuclear correlation spectrum (HMBC) (Figure 3.12). The important correlations arising from $^3J_{H,C,C}$ and $^3J_{H,O,C}$ are summarized in Table 3.6. Correlations of each of the anomeric resonances at 4.47, 4.42 and 4.38 ppm to the carbons across the glycosidic linkage (C-3 at 86.4 and the two C-4s at 79.5 and 79.3 ppm, respectively) define which anomeric proton can be assigned to structures **A** and **C**. Thus, the anomeric proton at 4.47 is assigned to H-1 of the (1→4)-linked β-D-glucose residue (**A**) that is glycosidically linked to the 3-position of residue **B** and attached by C-4 to another (1→4)-linked residue. Similarly, the anomeric proton at 4.38 ppm is assigned to H-1 of the (1→4)-linked β-D-glucose residue (**C**) that is linked through C-4 to the C-1 position of the (1→3)-linked β-D-glucose (**B**), as shown in Scheme 1. The concomitant

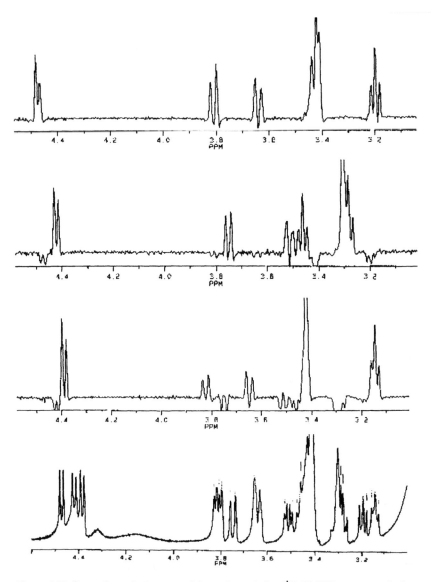

**Figure 3.11** Comparison of subspectra of the total correlation ($^1$H-TOCSY) spectrum of wheat β-D-glucan in deuterated dimethyl sulfoxide. Each spectrum (a, b and c) is a slice of the two-dimensional spectrum at the chemical shifts of the anomeric protons, giving the intra-residue proton shifts: (a) 4.47 ppm, corresponding to residue A in scheme 1; (b) 4.42 ppm, corresponding to residue B; (c) 4.38 ppm, corresponding to residue C; (d) the single pulse $^1$H spectrum of wheat β-D-glucan. Reprinted from Reference [41] with permission from Elsevier Science.

TABLE 3.5. Complete Assignment of $^{13}$C and $^1$H NMR Spectra of Wheat β-D-Glucan Based on Heteronuclear Correlation and Shift-Correlated Spectroscopy (COSY).

| Glucose Residue | Assigned C, H Position | $^{13}$C Resonance | $^1$H Resonance |
|---|---|---|---|
| (1→4)-Linked β-D-glucose | 1 | 103 | 4·47 |
| | 2 | 73·2 | 3·19 |
| (residue A in Scheme 1) | 3 | 74 | 3·43 |
| | 4 | 79·3 | 3·42 |
| | 5 | 74·7 | 3·42 |
| | 6 | 60·1 | 3·64 |
| | | | 3·81 |
| (1→3)-Linked β-D-glucose | 1 | 101·9 | 4·42 |
| | 2 | 71·9 | 3·28 |
| (residue B in Scheme 1) | 3 | 86·4 | 3·46 |
| | 4 | 68 | 3·3 |
| | 5 | 75·9 | 3·3 |
| | 6 | 60·5 | 3·51 |
| | | | 3·75 |
| (1→4)-Linked β-D-glucose | 1 | 102·1 | 4·38 |
| | 2 | 72·7 | 3·14 |
| (residue C in Scheme 1) | 3 | 74·3 | 3·42 |
| | 4 | 79·5 | 3·42 |
| | 5 | 74·6 | 3·42 |
| | 6 | 60·1 | 3·64 |
| | | | 3·82 |
| (1→4)-Linked β-D-glucose | 1 | 102·1 | 4·38 |
| | 2 | 72·7 | 3·14 |
| (residue D in Scheme 1) | 3 | 74·3 | 3·42 |
| | 4 | 79·2 | 3·42 |
| | 5 | 74·7 | 3·42 |
| | 6 | 60·1 | 3·64 |
| | | | 3·81 |

carbon to proton correlations in the reverse direction, i.e., 103.3 ppm (C-1 of **A**) to 3.46 ppm (H-3 of **B**) and C-1 of **B** and **C** to H-4 of **C** and **A**, respectively), are also observed but are not useful due to overlap of the H-4s from **A** and **C**. This, along with the additional intraresidue correlations observed between the resolved H-2s of each residue and their respective C-1 and C-3 resonances serves to confirm the assignments within each glucan residue and resolves any ambiguity between the C-4 assignments for **A** and **C**. In addition, coupling

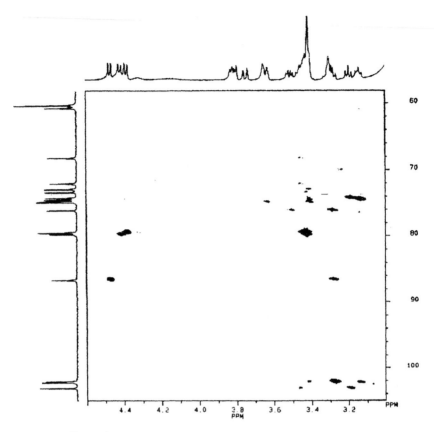

**Figure 3.12** $^{13}$C and $^{1}$H long-range heteronuclear correlation NMR spectrum (HMBC) of wheat β-D-glucan in deuterated methyl sulfoxide. The single pulse spectra are shown as the projections. Reprinted from Reference [41] with permission from Elsevier Science.

between H-3 and C-4 and H-4 and C-5 within residue **B**, confirmed the relative assignments of C-4 vs. C-5.

The difference between consecutive sequences of the trisaccharide and tetrasaccharide in this glucan can only be detected in the NMR spectrum through identification of residue **D**, the (1→4) linked β-D-glucose that is flanked by two (1→4) β-D-glucose residues. This β-D-glucose behaves much like cellulosic residues, and previous studies [14,40] have shown that the C-4 carbon of this residue can be assigned to the small resonance at 79.2 ppm, a shoulder on the resonance assigned to C-4 of **A** and **C**. The rest of the carbons and protons for this residue are too close to those of residues **A** and **C** to be resolved (Table 3.5) [41]. Comparison to the previously observed $^{13}$C spectrum for the cellodextrin-like residue of ~DP 9, which gives resonances at 102.1 (C-1), 79.3 (C-4),

**TABLE 3.6.** Linkage Sequences and Connectivities of Wheat β-D-Glucan Derived from Long-Range Correlation (HMBC) Spectra.[a]

| Resonance | Correlated Resonance |
|---|---|
| 4.47 ppm (H-1 of **A**) | 86.4 ppm (C-3 of **B**) |
| 4.42 ppm (H-1 of **B**) | 79.5 ppm (C-4 of **C**) |
| 4.38 ppm (H-1 of **C**) | 79.3 and 79.2 ppm (C-4 of **A** and **D**, respectively) |
| 103.3 ppm (C-1 of **A**) | 3.46 ppm (H-3 of **B**) |
| 101.9, 102.1 ppm (C-1 of **B** and **C**) | 3.42 ppm (H-4 of **A** and **C**) |
| 3.19 ppm (H-2 of **A**) | 74.0 ppm (C-3 of **A**) and 103.0 ppm (C-1 of **A**) |
| 3.14 ppm ( H-2 of **C**) | 74.3 ppm (C-3 of **C**) and 102.1 ppm ( C-1 of **C**) |
| 3.28 ppm (H-2 of **B**) | 86.4 ppm (C-3 of **B**) and 101.9 ppm (C-1 of **B**) |
| 68.0 ppm (C-4 of **B**) | 3.46 ppm (H-3 of **B**) |
| 75.9 ppm (C-5 of **B**) | 3.30 ppm (H-4 of **B**) |

[a]**A**, **B**, **C** and **D** are the sugar residues as described in Scheme 1. **A**: is the (1→4)-β-D-glucose residue glycosidically linked to the 3 position of the (1→3)-β-D-glucose (**B**) and attached through C-4 to a (1→4)-β-D-glucose. **B**: is the (1→3)-β-D-glucose residue. **C**: is the (1→4)-β-D-glucose residue glycosidically linked to the 4 position of the (1→4)-β-D-glucose (**A** or **D**) and attached by a (1→3)-β-D-glucose at the C-4 position. **D**: is the (1→4)-β-D-glucose residue flanked on either side by (1→4)-β-D-glucose residues.
Reprinted from Reference [41] with permission from Elsevier Science.

74.7 (C-5), 74.3 (C-3), 72.8 (C-2) and 60.2 (C-1), provides an excellent match [14,40].

The complete assignments of $^{13}$C and $^{1}$H spectra of cereal β-D-glucans using coupling relationships are in complete agreement with the original assignments of Dais and Perlin [40], especially those for residue **B**, which were based on chemical shift arguments and comparison to tetrasaccharides released by hydrolysis. However, these data provide a more detailed assignment of the (1→4)-linked β-D-glucan residues and sequences of the linkage. The resolution of the anomeric protons in the $^{1}$H spectrum may also provide a method for quantification of linkages.

## 3.4. IDENTIFICATION AND QUANTITATIVE ANALYSIS OF β-D-GLUCANS

### 3.4.1. Specific Dye Binding and Fluorescence Detection

Cereal β-D-glucans can interact specifically with two dyes, Calcofluor and Congo Red, respectively, to produce precipitation, increase in viscosity and bathochromic (red) shifts in the absorption spectra and increase in fluorescence intensity of the dyes [30,33,42,43]. Studies of complex formation between dye and β-D-glucans in solution, and of staining behavior, have provided a number

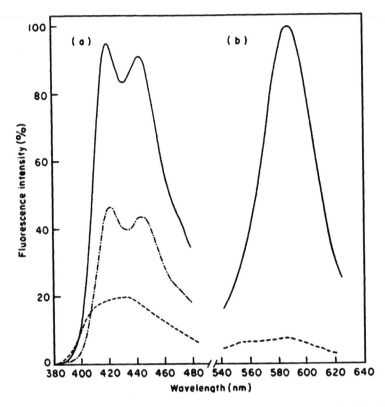

**Figure 3.13** Fluorescence emission spectra in phosphate buffer pH 7.0 ($I = 0.05$) of (a) Calcofluor (excitation 350 nm) and (b) Congo Red (excitation 470 nm) alone and in the presence of β-D-glucans. Dye alone (---); dye plus oat β-D-glucan (500 ug/mL) (—); Calcofluor plus alkaline extract from wheat aleurone cell walls (-·--·-). Reprinted from Reference [43] with permission from the *Journal of Cereal Science.* © Academic Press Ltd.

of methods of isolation, identification, localization and quantification of β-D-glucans in cereals [33]. The interactions between the two dyes and β-D-glucans are specific in oat, barley, rye and wheat. However, the presence of cellulose, xyloglucan or any other polysaccharides that also interacts with the dyes will interfere with the analysis. The effect of β-D-glucans on the absorption and fluorescence spectra of Congo Red and Calcofluor are shown in Figure 3.13 and Table 3.7. Oat β-D-glucan-induced red shifts with Congo Red increased with β-D-glucan concentration, from 10–12 nm at 2 μg/mL to an approximately constant shift of 25 nm at higher concentrations (greater than 10 μg/mL), and with an increase of fluorescence intensity of 12–13-fold [43]. The $\Delta\lambda_{max}$ of Calcofluor increased to a maximum of 12–14 nm at 50–100 μg/mL of oat β-D-glucans, and the maximum fluorescence intensities increased 4–5 times. Such

TABLE 3.7. Effects of Soluble Extracts from Cell Wall Fractions, Pentosan and β-Glucan Preparations on the Absorption Spectra of Congo Red and Calcofluor and on the Fluorescence Emission Spectra of Calcofluor.

| Preparation | Absorption $\Delta\lambda_{max}$ (nm)[a] | | Fluorescence Emission | |
|---|---|---|---|---|
| | Congo Red | Calcofluor | $\Delta\lambda_{max}$ (nm) Calcofluor | RFI[b] Calcofluor |
| 1. β-D-glucan from oats | 25 | 14 | 420,442 | 4.5 |
| 2. Alkali-soluble extract from oats | 7 | 13 | 420,442 | 4.3 |
| 3. Alkali-soluble extract from barley aleurone cell walls | 29 | 14 | 420,443 | 3 |
| 4. Alkali-soluble extract from wheat aleurone cell walls | 33 | 14 | 420,443 | 0.5 |
| 5. Water-soluble pentosan from wheat endosperm cell walls | 3 | −2 | 421,440 | 0.9 |
| 6. Pentosan from wheat flour | 14 | 3 | 421,441 | 1.3 |
| 7. Arabinogalactan-peptide from wheat | 3 | 1 | 420,436 | 0.3 |

[a] $\lambda_{max}$ of dye + sample minus $\lambda_{max}$ of dye.
[b] Relative fluorescence intensity: ratio of fluorescence intensity of Calcofluor + sample to that of Calcofluor alone.
Reprinted from Reference [43] with permission from the *Journal of Cereal Science.* © 1983 Academic Press Ltd.

substantial changes in absorption and fluorescence values have been used to quantitatively determine the content of β-D-glucans in raw material and food products, as well as to detect β-D-glucans after size exclusion chromatography [17,44]. Quantitative analysis of β-D-glucans is based on the linear relationship between the absorbance and β-D-glucan concentration at a fixed wavelength and dye concentration [45].

### 3.4.1.1. Flow Injection Analysis (FIA)

An automatic FIA apparatus can be made as described in Figure 3.14 [44]. The reagent [0.0025%, (w/v) Calcofluor in 0.1 M phosphate buffer, pH 7.0] and the carrier stream (0.1 M phosphate buffer, pH 7.0) are delivered at a flow rate of 2.0 mL/min. 10 μL of sample solutions are injected in the carrier stream

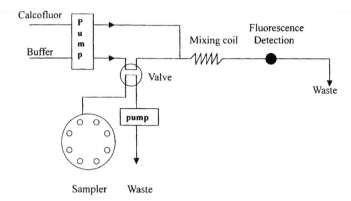

**Figure 3.14** Diagram of an apparatus used for the β-D-glucan analyzer based on FIA. Pump: peristaltic pumps; valve: Rheodyne, model 5011 p (Cotati, California, USA); detector: HPLC fluorescence detector, model RF-530 (Shimadzu, Kyoto, Japan); sampler: model AS-40 (Jasco, Tokio, Japan); tubings: 0.5 mm internal diameter PVC or Teflon tubing (mixing coil: 1.5 m). A timer to control the time cycles and a recorder are not shown. Reprinted from Reference [44] with permission. © 1985 European Brewery Convention.

and mixed with the Calcofluor solution. The increase in fluorescence can be observed at 360 nm excitation and 425 nm emission, respectively. The contents of β-D-glucans are derived from a standard curve (β-D-glucan standards are commercially available). The relationship between the amount of β-D-glucan and the increase in fluorescence intensity and a calibration line obtained using four different β-D-glucan preparations are shown in Figure 3.15(a) and (b), respectively. This is a rapid and sensitive automatic method for measuring β-D-glucans in raw materials and food products. However, the limitation of this

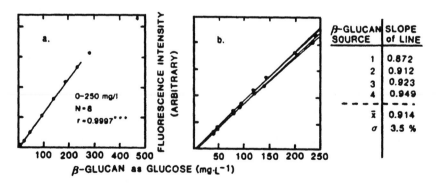

**Figure 3.15** (a) Relationship between the amount of β-D-glucan and the increase in fluorescence intensity using the device described in Figure 3.14. (b) Calibration lines determined using four different β-D-glucan preparations. Reprinted from Reference [44] with permission. © 1985 European Brewery Convention.

method is that it can only determine β-D-glucans that are soluble in water or buffers. β-D-glucans that are not soluble in aqueous solvents cannot be detected.

### 3.4.1.2. Fluorescence Detection

The dye-binding property of β-D-glucans can also be used as an on-line detection method of High-Performance Size Exclusion Chromatography (HPSEC). Wood et al. used a post-column dye-binding method to detect the chromatographic responses [17]. Calcofluor (0.01%, w/v) in Tris buffer at pH 8.0 (tris-hydroxymethy-aminomethane), protected from light, was introduced into the post-column stream through a T-junction at 0.5 mL/min and pumped through a 50 ft × 0.007 in. pulse-damping coil. Mixing is effected in a 10 ft × 0.01 in coil prior to measurement of fluorescence intensity in Spectrofluorimeter: the excitation and emission rates are set at 360 and 425 nm, respectively [17]. Typical size exclusion chromatographic profiles of β-D-glucans from oat and barley detected by fluorescence responses are demonstrated in Figure 3.16.

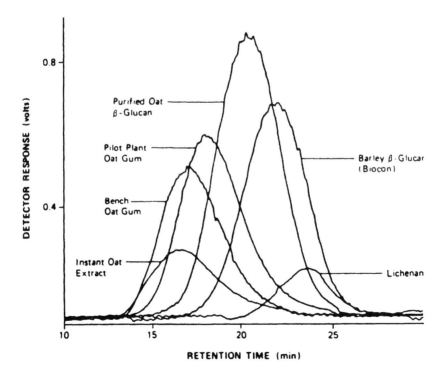

**Figure 3.16** Post-column fluorescence responses of cereal β-D-glucans from high-performance size-exclusion chromatography at flow rate 0.5 mL/min. Reprinted from Reference [33] with permission from The American Association of Cereal Chemists, © 1986.

### 3.4.1.3. Fluorescence Microscopy-Histochemistry

Specific dye-binding of β-D-glucans has also been used in the histochemical studies of β-D-glucans in cereal grains. Cell wall and carbohydrate fractions are processed through ethanol, *n*-propanol prior to embedding in glycol methacrylates [30]. Polymerized samples are sectioned as consistently and uniformly as possible (approximately 2 μm thick) using glass knives, and the sections are fixed to glass slides. The sections are stained with 0.01% Calcofluor (60 s) or Congo Red (5 min) in sodium phosphate buffer (pH 7.8), I 0.2 in 50% ethanol. Excess stain is washed with running water for 1 min, and the sections are air-dried, mounted under a cover glass in fluorescence-free immersion oil. Once sections have been mounted on the glass slides, they may be examined without further treatment using fluorescence microscopy to obtain the micrographs. Figure 3.17 shows the fluorescence micrograph of the aleurone and endosperm layers of oat and barley [46].

### 3.4.2. Enzymatic Method

A simple and quantitative enzymatic method for determining β-D-glucans was first developed for barley and malt samples by McCleary and coworkers

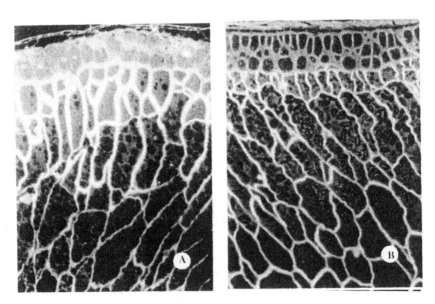

**Figure 3.17** (a) Transverse section of oat (cv. Hinoat) endosperm showing Calcofluor stained cell walls; the cell walls in the subaleurone region are of increased thickness. (b) Transverse section of barley (cv. Vanier) endosperm, showing Calcofluor stained cell walls.

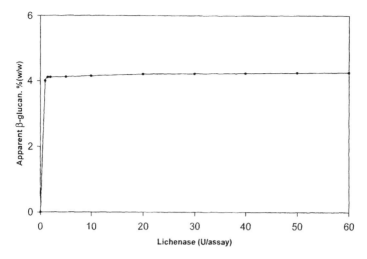

**Figure 3.18** The level of lichenase required in the standard assay procedure for β-glucan in barley flour. The arrow shows the amount of enzyme routinely used. Reprinted from Reference [48] with permission from the *Journal of the Institute of Brewers.* ©️ 1985 Institute of Brewers.

[47,48]. This method is extended to direct and complete analysis of β-D-glucans from other cereals and cereal products, such as oat and wheat. β-D-glucans are specifically depolymerized with a highly purified $(1 \rightarrow 3)(1 \rightarrow 4)$-β-D-glucanase (lichenase) from *Bacillus subtilis* to tri-, tetra- and higher DP of oligosaccharides (see section 3.3.2). These oligosaccharides are then further hydrolyzed to glucose using purified β-D-glucosidase. The released glucose is determined by glucose oxidase/peroxidase reagent to give absorbance at 510 nm [48]. The level of lichenase required in the standard assay is 10 unit/0.5 g of a flour sample (Figure 3.18). Standard glucose is used to make the calibration curve for quantification. A streamlined enzymatic procedure was described by McCleary and Codd [49], which is now used as a routine operation in our laboratory. In the standard assay procedure, oven-dried flour is suspended in phosphate buffer (pH 6.5) and incubated in a steam bath or boiling water (100°C) for 5 min to ensure complete inactivation of endogenous enzymes and hydration of the β-D-glucans. The system is treated with lichenase to depolymerize the glucan and is then adjusted to a set volume. This solution is filtered, and aliquots (0.1 mL) of the filtrate are treated with β-glucosidase. The released D-glucose is quantitated using glucose-oxidase/peroxidase reagent, and the change of absorbance at 510 nm is converted to the amount of β-D-glucans. The presence of free glucose or glucose from the degradation of starch will interfere with the analysis. The presence of free D-glucose can be corrected by using blank samples without adding lichenase and β-D-glucosidase. β-D-glucans in oven-dried flour samples were readily accessible to lichenase hydrolysis. The oven-drying process

inactivated most, but not all, of the endogenous enzyme activity that might interfere with the assay. The remaining enzymatic activities are inactivated during the "cooking" step (100°C, 5 min). The level of lichenase required to give quantitative hydrolysis of β-D-glucans to oligosaccharides was determined using varying levels of lichenase. The result obtained was expressed as β-D-glucan content after treatment of the lichenase-released oligomers with β-glucosidase followed by D-glucose determination. A lichenase of two units per assay is usually adequate; however, to ensure the complete hydrolysis, 10 units are used in the standard procedure [48]. As discussed in section 3.3.2, an insoluble white precipitate was formed after hydrolysis of pure β-D-glucans by lichenase. This material has a higher d.p. composed of consecutive (1→4)-β-linked D-glucosyl residues with a 3-linked D-glucosyl residue at the reducing terminal. This precipitate represents about 3% of total water-soluble β-D-glucan, but it may also represent a higher proportion of the insoluble fraction. As a result, this method may lead to a slight but significant underestimation of the total β-D-glucan content as the precipitate is removed by the filtration step. Nevertheless, this method is still considered to be more accurate than the dye-binding method. The dye-binding method can only determine the amount of β-D-glucans in aqueous solutions; while the enzymatic method is able to break down soluble and insoluble β-D-glucans.

## 4. PHYSICAL PROPERTIES: SOLUTION, CONFORMATION AND RHEOLOGICAL PROPERTIES

### 4.1. MOLECULAR WEIGHT, CONFORMATION AND SOLUTION PROPERTIES OF β-D-GLUCANS

#### 4.1.1. Molecular Weight and Molecular Weight Distribution

There are two major fundamental properties of polysaccharides that determine the overall functionality of the macromolecules: one is primary structure, as described in section 3.3; the other is molecular weight and molecular weight distribution. Molecular weight and MW distribution affect the overall functionality of polysaccharides by affecting their hydration and conformation properties in aqueous systems, which ultimately, will affect the viscosity and rheological behavior of the system. Because of the complexity and polydispersity of natural polysaccharides, there is no simple way to accurately express their molecular weight. In polymer science, different average molecular weights are obtained depending on the method used. Number-average molecular weight, $M_n$, is simply the total weight of substance divided by the total number of moles present:

$$M_n = \sum M_i N_i \Big/ \sum N_i \qquad (3\text{-}1)$$

Mn is determined by the colligative properties of solutions (e.g., osmometric method). In this method, each molecule, large or small, makes the same contribution to the observed effect. Some properties, for a given concentration, are proportional to the molecular weight of the solute. The effect obtained from a polydispersed molecule, such as a polysaccharide, will depend on its weight average molecular weight, Mw:

$$Mw = \sum NiMi^2 \Big/ \sum NiMi = \sum CiMi^2 \Big/ \sum CiMi \qquad (3\text{-}2)$$

A number of methods can be used to determine the weight average molecular weight, including light scattering, sedimentation, diffusion and flow birefringence. From the above two equations, it can be seen that large molecular species are weighted more heavily in the weight average than the number average. Therefore, for any polydispersed material, including natural polysaccharides, Mw > Mn. The ratio Mw/Mn is used to characterize the polydispersity of a polymer. Another frequently used average molecular weight is z-average molecular weight, Mz:

$$Mz = \sum NiMi^3 \Big/ \sum NiMi^2 \qquad (3\text{-}3)$$

For natural polysaccharides, the three average molecular weights follow the order of Mz > Mw > Mn.

The most important absolute methods for determining the average molecular weight of polysaccharides are light scattering (Mw), membrane osmometry (Mn) and sedimentation (Mw) analysis. However, these methods do not give information on molecular weight distribution if using a single method. Size exclusion chromatography potentially provides a simple method for determining MW and MW distribution, and it is fast, simple and reproducible compared to classical methods. Unfortunately, this method has two major drawbacks. First, there is currently a lack of filling materials with pore sizes large enough to give good fractionation in the high molecular weight range. Second, since the fractionation occurs according to molecular size (hydrodynamic volume) rather than real molecular weight, the columns have to be calibrated with standards of the same flexibility (ratio of contour length to persistence length) as the polysaccharide under investigation or some universal calibration procedure. Commercially available dextran and pullulan standards have often been used in the literature despite the fact that they may lead to overestimation of the molecular weight of β-D-glucans because of the differences in molecular conformation (shape) between the α-D-glucans and β-D-glucans [50]. The MW values obtained by this method should be reported as dextran or pullulan equivalents [38,50,51]. The product of intrinsic viscosity and molecular weight ([η] × M), which is proportional to the polymer hydrodynamic volume, has been demonstrated as an universal calibration parameter for some synthetic polymers. When the Mark-Houwink-Sakurada (MHS) parameters for the calibration standards

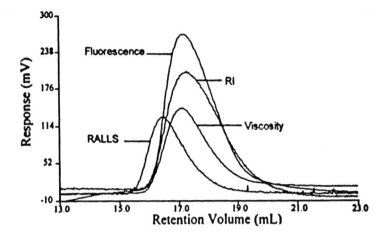

**Figure 3.19** High-performance size-exclusion chromatogram of an oat β-glucan standard (J271) using refractive index (RI), viscometry, right-angle laser light-scattering (RALLS) and fluorescence detection. Reprinted from Reference [21].

and the polysaccharide under investigation are known, the molecular weight of the standards may be converted into the molecular weight of the unknown polysaccharides that would elute in the same volume. Vårum and colleagues [38,50,52] reported that the MW of oat β-D-glucan followed the following Mark-Houwink-Sakurada relationship [38,50]:

$$[\eta](mL/g) = 6.7 \times 10^{-2} \, Mn^{0.75} \qquad (3\text{-}4)$$

Here, Mn was determined by the osmometric method.

Recently, size exclusion chromatography systems using a light-scattering detector in combination with a refractive index and viscometric detectors have become commercially available. The combination of size exclusion chromatographic profiles and the result calculated from multidetectors allows simultaneous characterization of molecular weight/shape and molecular weight distribution. Figure 3.19 demonstrates the responses of four different detectors to one β-glucan sample. The light-scattering detector tends to give a lower elution volume than the other three detectors. This is because the light-scattering detector is much more sensitive to large molecules versus small molecules. In crude extracts of oat and barley, the use of these nonspecific detectors to evaluate β-glucan is not possible because of unknown contributions from co-extracted substances (arabinoxylans, starches and proteins). This problem can be overcome by using a fluorescence detector exploiting the specific dye binding of Calcofluor by β-glucan [18]. Typical size exclusion chromatograph profiles of different β-glucans using on-line Calcoflour dye binding and fluorescence detection are shown in Figure 3.16. This system may also be used to quantitate β-glucan [18,53].

Using high-performance size-exclusion chromatography with Calcofluor detection and oat β-glucan standards precalibrated by LALLS (low-angle laser light scattering), Wood et al. [17] reported MW of β-glucan in crude extracts from several oat, barley and rye cultivars. The MW of oat β-glucan ($\sim 3 \times 10^6$) was greater than that from barley (2-2.5 $\times 10^6$) or rye ($1 \times 10^6$). Bhatty also suggested that the MW of oat and barley β-glucan is greater than $2 \times 10^6$, and the MW of oat is greater than that of barley [54]. Mälkki et al. reported lower values ($1.5 \times 10^6$) for β-glucan from oat bran concentrates during an examination of differences in MW of β-glucans related to processing conditions [38]. Using columns calibrated with β-glucan standards with MW that were pre-established using RALLS (right-angle laser light scattering) and viscometric methods, Beer et al. reported somewhat lower MWs of β-glucans in extracts from various oat and barley cultivars than those reported by Wood et al. [17,21], but they confirmed a significant difference between oats and barley. Autio et al. [39] reported cultivar variations in MW with a highest value of $1.5 \times 10^6$. These data are summarized in Table 3.8 [15]. Discrepancies can be

TABLE 3.8. **Comparison of Molecular Weights of β-Glucan from Cereals.**

| Source | Pt[a] | Extraction Solvent and Condition | | | Detection Method[b] | MW ($\times 10^{-6}$) |
| | | Solvent | T°C | Duration (min) | | |
|---|---|---|---|---|---|---|
| Wheat bran (preprocessed) | 2 | 1 M NaOH | 25 | 120 | B,C | 0.3–0.6 |
| Wheat bran (AACC, soft white) | 2 | 1 M NaOH | 25 | 120 | B,C | 0.4–0.7 |
| Wheat bran (AACC, red hard) | 2 | 1 M NaOH | 25 | 120 | B,C | 0.27–0.6 |
| Barley flour | 2 | $Na_2CO_3$ | 45 | 90 | B,C | 0.8 |
| Barley flour | 2 | $H_2O$ | 65 | 120 | A,C | 0.6 |
| Barley cell wall (endosperm) | 2 | $H_2O$ | 40–65 | — | D | $\sim 40$ |
| Oat bran | 1 | $H_2O$ | 70 | 60 | A | 1.5 |
| Oat flour | 2 | $H_2O$ | 90 | 120 | B,C | 2.0–2.5 |
| Oat bran | 1 | $Na_2HPO_4$(20 mM) | 37 | 135 | B,C | 1.1–1.9 |
| Oat bran | 1 | $Na_2CO_3$ | 70 | 120 | A | 0.4–1.5 |
| Oat bran | 2 | $Na_2CO_3$ | 60 | 120 | B,C | 2.7 |
| Flour | 2 | $Na_2CO_3$ | 60 | 120 | B,C | 3.0 |
| Oat bran | 2 | $Na_2CO_3$ | 60 | 120 | B,C | 3.1 |

[a]Sample pretreatment: 1 = no pretreatment; 2 = sample treated with hot aqueous or ethanol prior to extraction.
[b]Detection Method: A = multi-angle laser light-scattering detection; B = Calcofluor post-column detection with β-glucan standards; C = refractive index detection with pullulan standards; D = sugar concentration.
Reprinted from Reference [15] with permission.

attributed to the measuring system and extraction and isolation methods. It is unlikely that a "true" MW or MW distribution of oat β-glucan as it exists in the cell wall is measurable. It is, however, possible to aim for determining the maximum possible MW that can be extracted.

The molecular weight of wheat β-glucan is in the range of 300,000 to 600,000. Isolated and purified wheat β-glucan is in the lower molecular weight range, while the direct measurement of crude extract gives a higher molecular weight [12,28] (Table 3.8).

Sample pretreatment prior to extraction is important to determining the molecular weight of the isolated β-glucan. For example, if β-glucanases are present, they must be inactivated. Wood et al. [17] reported that, without inactivation of β-D-glucanases, β-D-glucan extracted from commercial oats was of lower MW compared to β-glucan from oats pretreated with ethanol (2 hr at 85°C, 70% ethanol) or steam-heat treated. The effects of different heat treatments of the grain on the relative viscosities of extracted β-D-glucans are shown in Table 3.9 [55]. β-glucans from raw material without any treatment give the lowest relative viscosity, while β-glucans from steamed grain had the highest relative viscosity although the extraction rate was the lowest. Their results indicate that steaming is a more effective way of deactivating the enzymes compared to roasting.

The measurement of MW requires material to have been properly dispersed, ideally molecularly. Vårum et al. reported an aggregation phenomenon for depolymerized oat β-glucan [52]. It has also been observed that wheat and barley

TABLE 3.9. **Effect of Genotype and Heat Treatment of Whole Oats on the Relative Viscosity[a] of Water Soluble Extracts, the Concentrations of Soluble and Total β-Glucan and Ratio of Soluble and Insoluble β-Glucan in Oat Flour.[b]**

|  | Relative Viscosity | Soluble β-Glucan (g kg$^{-1}$) | Total β-Glucan (g kg$^{-1}$) | Soluble/Total β-Glucan Ratio |
|---|---|---|---|---|
| **Treatment** | | | | |
| Raw | 2.18 | 39.8 | 48.3 | 0.82 |
| Roasted | 2.27 | 40.0 | 47.2 | 0.84 |
| Steamed | 3.12 | 34.4 | 46.7 | 0.74 |
| **Genotype** | | | | |
| Marion | 2.91 | 46.0 | 55.5 | 0.83 |
| Steele | 2.45 | 40.1 | 47.1 | 0.85 |
| Robert | 2.21 | 28.0 | 39.7 | 0.71 |

[a]Cannon Ubbellonde semi-micro capillary viscometer, size 75.
[b]Effects of treatments are reported as means of all genotypes; genotype effects are reported as means of all treatments, $n = 6$

Reprinted from Reference [55] with permission from the *Journal of the Science of Food and Agriculture.* ©1997 John Wiley & Sons Limited.

β-glucans can form gels under storage conditions (4°C–25°C) [12,37,56]. Lower temperature and repeated cycles of short-term steady shear and rest periods accelerated the sol-gel transition [12,37,56]. Therefore, heating of the sample before HPSEC (high-performance size exclusion chromotography) analysis is necessary. A column temperature of 77.5°C was used by Bohm and Kulicke [37,56] in order to avoid aggregations.

As discussed in section 2.1, extraction conditions will affect the molecular weight of the extracted β-D-glucans. Due to the differences in solubility and stability of β-glucans in various solvents, decisions need to be made about whether to use mild conditions and get less extraction or more vigorous reagents and risk degradation. Using hot water or carbonate buffer (pH 10) as extraction media [18,21], only 50–70% of total β-glucan was solubilized, leaving over 30% of the total β-glucan insoluble and of unknown MW. Sodium hydroxide appeared to give total extraction, but the work of Beer et al. showed that the MW of β-glucan was significantly decreased by NaOH [21,22].

McCleary [18] reported that increasing the temperature of extraction leads to an increase in the MW of barley β-glucans as determined by the intrinsic viscosity of extracts. Similarly, Beer and coworkers found that the MW of oat β-glucan extracted at 90°C was up to 30% higher than that of β-glucan extracted at 37°C [29]. These observations suggest that high molecular weight species are more soluble and, therefore, extractable at high temperatures.

## 4.1.2. Conformation and Solution Properties

Similar to many other polysaccharides, β-D-glucans exist in solution as conformationally disordered "random coils" whose shapes fluctuate continually under Brownian motion [57]. At low concentrations, the individual coils are separated from each other and are free to move independently. When concentration is increased to a certain stage, those independent random coils start to touch, overlap and interpenetrate one another. Accompanying the relative status of the coils, is their solution properties, such as solution viscosity. The concentration dependence of limiting specific viscosity of oat β-glucans is shown in Figure 3.20 [58]. The transition from a dilute solution of independently moving coils to an entangled network is evidenced by the dramatic change of the slope from 1.08 to 3.90.

The conformation of β-D-glucan chains also varies significantly in different solvent systems. These variations in conformation will affect the hydrodynamic volume, hence, the molecular weight is determined based on it. As shown in Table 3.10 [59], the measured Mw of a β-D-glucan sample isolated from beer has the highest Mw, radius of gyration and intrinsic viscosity in water, followed by in 90% DMSO, 2M GHCl, Cadoxen and Cuoxam I and II. The metal complexing solvents, Cuoxam and Cadoxen, which are known to be good solvents for cellulose, are able to minimize the interactions between β-D-glucan molecules. These solvents dissociated the aggregates and thus allowed determination of the

**TABLE 3.10. Molecular Characteristics of β-D-Glucans from Static Light-Scattering Measurement in Highly Diluted Solutions of Different Solvents.**

| Solvents | $c$ (mg/mL) | Mw (g/mol) | $x$ | $R_g$ (nm) | A2 (mol mL/g$_2$) | [η] (mL/g) |
|----------|-------------|------------|-----|------------|-------------------|------------|
| Water | 1.053 | $5400 \times 10^3$ | 30.5 | 197 | 0.0000413 | 219 |
| 90% DMSO | 0.9632 | $3050 \times 10^3$ | 17.2 | 165 | −0.00000875 | 180 |
| 2 M GHCL | 0.9782 | $2920 \times 10^3$ | 16.5 | 163 | −0.000162 | 138 |
| Cadoxen | 1.104 | $583 \times 10^3$ | 3.3 | 131 | 0.000454 | n/d |
| Cuoxam I | 1.0212 | $177 \times 10^3$ | 1.0 | 127 | 0.00101 | n/d |
| Cuoxam II | 0.968 | $170 \times 10^3$ | 0.96 | 117 | −0.00372 | n/d |

The temperature of measurement was always 20°C. C, concentration; Mw, molecular weight; $x$, degree of aggregation; $R_g$, radius of gyration; A2, second osmotic virial coefficient; [η], intrinsic viscosity.
Reprinted from Reference [59] with permission from Elsevier Science.

Mw of the unimer. In a double logarithmic plot of $R_g$ vs. apparent molar particle weight Mw, app [59] (Figure 3.21), a low exponent $a_{R_g} = 0.22$ was found:

$$R_g = K\,\text{Mw, app}^{0.22} \qquad (3\text{-}5)$$

The presence of maltose also exerted a strong and specific influence on the conformation of β-D-glucans, as shown in Figure 3.22. It appeared that

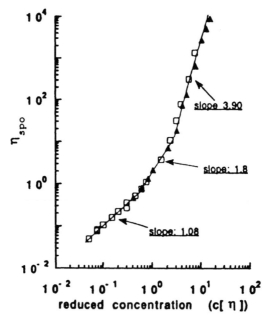

**Figure 3.20** Limiting specific viscosity (η$_{SDO}$) as a function of the reduced concentration ($c$[η]). OG0 (□), guar gum (▲). Reprinted from Reference [58].

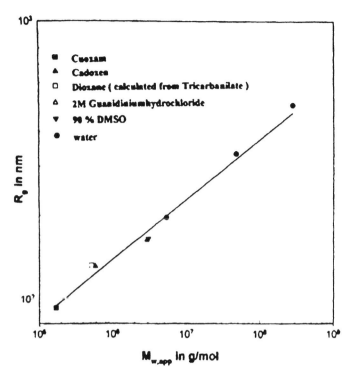

**Figure 3.21** Particle weight (M, app) dependence of the radius of gyration ($R_g$) of the β-glucan at 0.1% polysaccharide concentration in various aqueous and nonaqueous solvents. (●), water obtained by dilution from stock solutions of various concentrations; (Δ) 2 M guanidinium hydrochloride: (▼), DMSO/H$_2$O = 90/10 (v/v); (■), cuoxam; (▲), cadoxen. The observed dependence follows the relationship $R_g = KMw,app^{0.22}$ (this dependence describes the behavior of aggregates). Reprinted from Reference [59] with permission from Elsevier Science.

there is a particular maltose concentration (5%) at which both the Mw, app and intrinsic viscosity [η] exhibit the lowest value. Table 3.11 summarizes the effect of maltose concentration of the molecular characteristics of β-D-glucans isolated from beer [59]. The results of static light scattering seem not to follow the same $R_g$ vs. Mw dependence. Similar behavior was observed using the independent rheological method (Figure 3.23) which confirmed a change of structure at 5% maltose: a minimum of the reduced viscosity was also found in 5% maltose/water mixtures. The effect of maltose on the conformation of β-D-glucan is interpreted as a preferential binding of maltose to β-D-glucans that partially breaks up the aggregated clusters, and simultaneously, the cluster rigidity is decreased. The data for [η] and $R_g$ in pure water do not fall on the corresponding parabola of the maltose data.

The aggregation behavior of oat aleurone β-glucan is significantly different from that of barley β-glucan. For example, oat β-glucan can reach complete

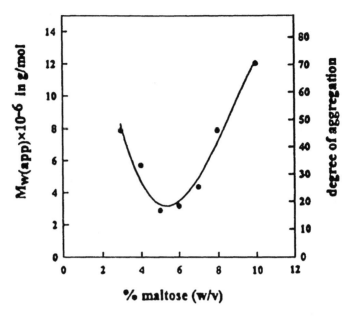

**Figure 3.22** Dependence of the apparent molar particle weight (Mw,app) on the aqueous maltose concentration. The degree of aggregation (X) is based on the molecular weight obtained in Cuoxam. Reprinted from Reference [59] with permission from Elsevier Science.

dissociation into monomers on dilution to zero concentration, but that is not true for barley β-glucan [50,59]. There was one report that oat aleurone β-glucan gave a pH of 4.06–4.85, indicating negative charges on the chain; however, no polyelectrolyte character was found with barley β-glucans. The charged polymer chain might exert repulsive interactions that reduce the aggregation

**TABLE 3.11. Molecular Characteristics of Highly Diluted β-Glucan in Aqueous Maltose Solution Measured at a Concentration of Approximately 1 mg/mL Solvent Mixtures.**

| Solvents | $c$ (mg/mL) | Mw (g/mol) | $X$ | $R_g$ (nm) | A2 (mol mL/g$_2$) | [η] (mL/g) |
|---|---|---|---|---|---|---|
| 2% Maltose | 1.372 | 8650 × 103 | 48.9 | 209 | $1.12 \times 10^{-4}$ | 281 |
| 3% Maltose | 1.048 | 7870 × 103 | 44.5 | 191 | $9.66 \times 10^{-5}$ | 275 |
| 4% Maltose | 1.004 | 5690 × 103 | 32.1 | 151 | $8.67 \times 10^{-5}$ | 200 |
| 5% Maltose | 1.02 | 2870 × 103 | 16.2 | 146 | $5.2 \times 10^{-5}$ | 129 |
| 6% Maltose | 0.996 | 3150 × 103 | 17.8 | 126 | $6.74 \times 10^{-5}$ | 186 |
| 7% Maltose | 0.948 | 4340 × 103 | 24.5 | 137 | $9.76 \times 10^{-5}$ | 225 |
| 8% Maltose | 0.984 | 7985 × 103 | 45.1 | 147 | $1.11 \times 10^{-4}$ | 401 |
| 10% Maltose | 0.803 | 12000 × 103 | 67.8 | 247 | $1.87 \times 10^{-4}$ | 466 |

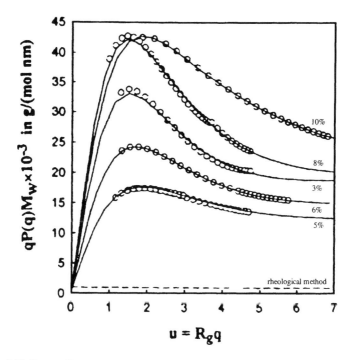

**Figure 3.23** Cassasa-Holtzer plot for 0.1% glucan solution dissolved in aqueous maltose solutions of the indicated percentage. The Cassasa-Holtzer plot allows determination of the chain stiffness of semiflexible chains applying a fitting program derived from the Koyama theory, $N_k$ is the number of statistically independent Kuhn segments per chain. Reprinted from Reference [59] with permission from Elsevier Science.

tendency and allow a complete dissociation on dilution. Grimm et al. concluded that the aggregation process was basically the same for both β-glucans; the main difference was probably due to the electrical charges that shift the equilibrium more toward dissociation [59]. In our own study, wheat β-glucan was not soluble in water, but the solubility increased significantly when pH decreased to 4 [14].

Gomez et al. [60,61] found that the chain of barley β-glucan could be satisfactorily modeled by a partially stiff worm-like cylinder with persistence length of 3.5–3.8 nm and cross-section diameter of 0.45 nm; this corresponds to an average of about four (1→3) β links per statistical segment. An investigation based on X-ray fiber diffraction and conformational analysis provided a conformation crystalline structure for lichenan and barley β-glucan [62]. The analysis of the fiber diagram of the hydrated form of lichenan led to a trigonal unit cell of dimension $a = b = 9.92$ Å and $c$ (fiber axis) $= 42.03$ Å, with cellotriose as the asymmetric unit and with six cellotriose residues per unit cell. A right-handed threefold helix made up of three cellotriose residues was derived from a comparison of experimental helical parameters with iso energy conformational maps. An antiparallel arrangement of the two chains was confirmed by refinement of geometrical and packing parameters [62]. Barley β-D-glucan crystallized in exactly the same fashion as lichenan, but the X-ray patterns displayed less order, i.e., greater line broadening. From the known structural features of β-D-glucans including lichenan, it is predicted that the crystalline structure of β-D-glucans (if there is any) will follow the increased order of oat, rye, barley, wheat and lichenan.

### 4.1.3. Solubility of Freeze-Dried β-D-Glucan Samples

Drying methods have a significant effect on the ease with which cereal β-D-glucans may be dispersed into solutions. For example, freeze-dried water-soluble barley β-D-glucan extracted at 65°C proved difficult to dissolve in aqueous buffer; it required several hours at 90°C with constant stirring before dissolution was completed [6]. In contrast, freeze-dried 40°C water-soluble barley β-glucan dissolved after 1–2 hr at 70°C. Freeze-dried wheat β-glucan is essentially insoluble in water regardless of heating temperature and time [18]. Past experience in our laboratory also showed that freeze-dried oat β-glucan is very difficult to dissolve in water. However, all cereal β-glucan samples prepared using the solvent exchange method in our laboratory exhibited good solubility in water. In other words, all samples prepared by the solvent exchange method are aqueous soluble with moderate heating.

### 4.2. RHEOLOGICAL PROPERTIES: STEADY FLOW

Cereal β-glucan is a linear, high molecular weight polysaccharide. The shear-rate ($\dot{\gamma}$) dependence of viscosity of β-D-glucans is typical of a "random coil" solution: there is a Newtonian plateau at the low shear-rate region and a shear thinning region at the higher shear-rate region, as shown in Figure 3.24 [58,63]. At low shear rates ($\dot{\gamma}$), the viscosity ($\eta$) of entangled "random coil" solutions remains constant at its maximum value ($\eta_0$) (the zero shear viscosity), because there is sufficient time for new entanglements to form between different chain partners to compensate for the disentanglement caused by shear flow. As a result, the overall "cross-link density" of the network remains constant, therefore, the viscosity remains unchanged. At higher shear rates ($\dot{\gamma}$), the rate of re-entanglements falls behind the rate of disruption of the existing entanglements, the overall "cross-link density" of the network decreases progressively with increasing shear rate and the viscosity falls (shear thinning). The onset shear-rate value at which the shear thinning flow behavior begins is concentration and molecular weight dependent, as shown in Figure 3.24. The flow curves of

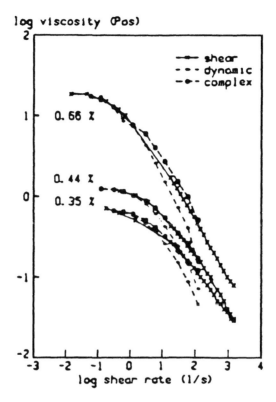

**Figure 3.24** Viscosities of β-glucan water solution. Reprinted from Reference [63] by permission of Oxford University Press, © 1988.

solutions of β-glucan can be described by the power law model:

$$\eta = K \dot{\gamma}^{n-1} \qquad (3\text{-}6)$$

where $\eta$ is viscosity (Pas), $K$ is consistency index (N.S$^n$/m$^2$), $\dot{\gamma}$ is shear rate (S$^{-1}$) and $n$ is flow behavior index. When $n = 1$, the viscosity is shear-rate independent, i.e., the Newtonian flow behavior; when $n < 1$, the viscosity deceased with the increase of shear rate, i.e., the shear thinning flow behavior. Changes of $n$ and $K$ of oat β-D-glucans with concentration in water and sugar solutions are summarized in Table 3.12 [63]. A linear relationship of the double log plot of the consistency index vs. concentration was also observed (Figure 3.25) [64]. The increase of β-D-glucan concentration increased the consistency index $K$ and decreased the flow behavior index $n$ indicating that the solution became more pseudoplastic as the concentration increased.

Figure 3.26 shows the concentration dependency of the relative viscosity of β-glucans with a series of molecular weights. It is evident that the viscosity of

TABLE 3.12. Power Law Constants for β-Glucans in Water and in 25% Sucrose Solution at 25°C.

| β-Glucan Powder Conc. (%, w/w) | $n_w^a$ | $n_s^b$ | $K_w^a$ (N.s$^n$/m$^2$) | $K_s^b$ |
|---|---|---|---|---|
| 0.20 | 0.84 | 0.83 | 0.04 | 0.06 |
| 0.45 | 0.62 | 0.53 | 0.57 | 1.49 |
| 0.67 | 0.50 | 0.40 | 3.40 | 7.28 |
| 0.89 | 0.30 | 0.29 | 12.2 | 21.3 |
| 1.00 | 0.28 | 0.26 | 24.2 | 31.2 |
| 1.10 | 0.24 | 0.23 | 36.1 | 40.3 |
| 1.22 | 0.20 | 0.21 | 46.9 | 50.7 |

[a] $n_w$ and $k_w$ are for water solution.
[b] $n_s$ and $k_s$ are for 25% sugar solution.
Reprinted from Reference [64] with permission from IFT.

higher molecular weight species increased faster than the low molecular weight β-glucans. The extreme examples are β-glucans with the highest and lowest weight average molecular weight (A, Mw = 573,000 and N, Mw = 9200). The relative viscosity of sample A increased sharply with concentration, while the relative viscosity of sample N is almost independent of its concentration. Viscosity of a polysaccharide solution is dependent on the volume the polysaccharide occupies in the solution. Higher molecular weight species have a much lower critical concentration ($c^*$) at which the inter-chain entanglements begin. At concentrations above $c^*$, the viscosity increases much more steeply with increase of concentration.

Since the viscosity ($\eta$) of β-D-glucan is concentration and shear-rate ($\dot{\gamma}$) dependent, any comparison of viscosity data must be made at the same shear rate and the same concentration. However, the consistency index $K$ and the flow behavior index $n$ of the power law model are shear-rate independent, therefore, a comparison of $K$ and $n$ values at the same concentration will be a valid approach. On this basis, Wood et al. [14] reported that oat gum (~80% β-D-glucan) prepared in the laboratory had higher $K$ values (theoretical viscosity at 1 s$^{-1}$) and lower $n$ values than pilot-plant-prepared oat gum at the same concentration. An oat gum prepared by Autio et al. [64] had similar properties to the laboratory prepared gum of Wood et al. [14]. Two samples prepared by Bhatty [54] using alkaline extraction exhibited significantly higher $n$ values than those reported by Wood et al. [14] and Autio et al. [64]. A survey by Autio et al. [39] indicated that different cultivars yield β-glucans of widely different values of $n$ (0.48–0.93), possibly indicating a wide range of MW variations. These measurements were done at concentrations (~0.3%) close to the dilute region where less molecular entanglement leads to more Newtonian behavior (e.g., $n$ = 0.93). Measurements reported by Wikstrom et al. [65] were also at

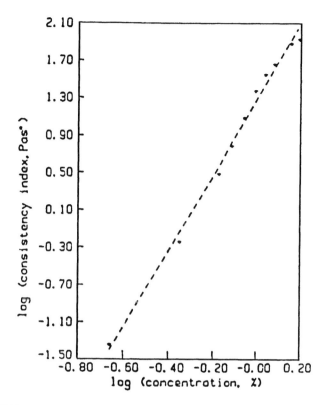

**Figure 3.25** Dependence of the consistent index *n* on β-D-glucan concentrations. Reprinted from Reference [61] with permission from Elsevier Science.

low concentrations but over a much wider shear-rate range, and *n* values ~0.45 were found that were indicative of greater shear thinning flow behavior. This illustrates the difficulty of making useful comparisons between data obtained from different laboratories. The difficulties were further illustrated by Wood et al. [9] where, depending on concentration and shear rate used for comparison, one oat gum preparation might appear 1.5 to 16 times more viscous than another.

According to Cox and Merz [66], shear and oscillation flow curves should be identical if the polymer solutions are devoid of energetic interactions. Weak-gel structures, such as xanthan gum and yellow mustard gum, deviate from this [67, 68]. Bohm and Kulicke showed that barley β-D-glucan BG300 had identical shear and oscillation flow curves at polymer concentrations of 2 and 4%. At higher concentrations (6 and 10%), shear viscosity decreases more rapidly than the complex viscosity in the non-Newtonian region, whereas zero shear and zero complex viscosity were identical, as shown in Figure 3.27. Autio et al.

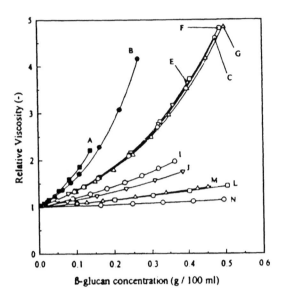

**Figure 3.26** Comparison of viscosity-concentration dependence of different molecular weight β-glucans in aqueous solutions. Reprinted from Reference [60] with permission from Elsevier Science.

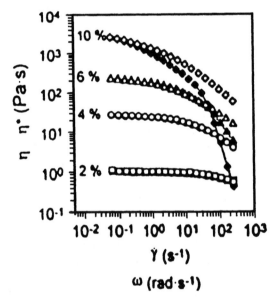

**Figure 3.27** Comparison of shear viscosity η as a function of shear rate (solid symbols) and complex viscosity η* as a function of angular frequency ω (open symbols) for a barley β-D-glucan at different concentrations. Reprinted from Reference [37] with permission from Elsevier Science.

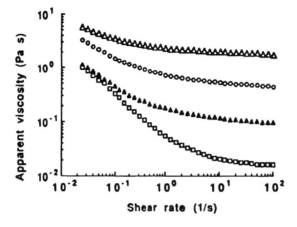

**Figure 3.28** Apparent viscosity as a function of shear rate of acid-hydrolyzed oat gum (OG60) solutions. Concentrations: 5.5% (△), 3.90% (○), 2.75% (▲), 1.15% (□). Reprinted from Reference [58].

also reported that β-D-glucans in 0.35% to 0.66% solutions obeyed the Cox-Merz rule at low shear rates; in contrast, the steady viscosity is higher than the dynamic viscosity at higher shear rates when the same numerical values of ω and $\dot{\gamma}$ are compared [63].

Small molecular weight cereal β-D-glucans exhibit some unusual flow behaviors. For example, the flow behavior index ($n$) of partially hydrolyzed oat β-D-glucans is smaller at lower concentrations compared to higher concentrations [58] (Figure 3.28). In other words, lower molecular weight oat β-D-glucans exhibit more shear thinning at lower concentrations than at higher concentrations; this is contradictory to higher molecular weight oat β-D-glucans [58,63]. Two low MW barley β-D-glucans (MW − 573,000, 0.16% and MW = 231,000, 0.31%, respectively) also exhibited profound shear thinning flow behavior in the shear-rate range 0.01–100 S⁻¹ (Figure 3.29) [63]. However, hot water extracted high Mw barley β-D-glucans (MW = 800,000) showed typical random-coil flow behavior. These unusual shear thinning rheological behaviors of degraded β-D-glucans are similar to rigid-rod types of polymers such as xanthan gum. The change of rheological flow behaviors suggests that the basic conformation of β-D-glucan has been changed from a random-coil type (in the high MW solutions) to a more ordered form in lower MW β-D-glucan solutions, perhaps caused by the associations of two or more β-D-glucan molecules in the three-dimensional network. This hypothesis is consistent with the fact that low MW β-D-glucans are able to form weak-gel structures and produce solid gels under cooling conditions, while the high MW samples are not, as described in the following two sections.

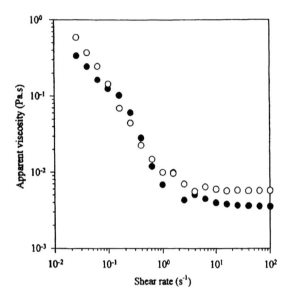

**Figure 3.29** Flow curves of barley β-glucan solutions of samples A (0.16%, black circles and G (0.31%, white circles). Temperature: 25°C. Reprinted from Reference [70] with permission from Elsevier Science.

## 4.3. RHEOLOGICAL PROPERTIES: VISCOELASTIC PROPERTIES

As discussed in the previous section, high MW cereal β-D-glucans exhibit similar flow behavior to that of random-coil polysaccharides. The mechanical spectrum of a β-D-glucan solution is also similar to that of random-coil polysaccharides, showing a typical concentrated solution pattern, as shown in Figure 3.30. The results of oscillatory measurements of Autio [63] were similar

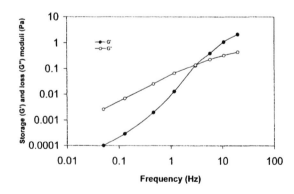

**Figure 3.30** Mechanical spectrum of β-D-glucan 2.0% (w/v) in aqueous solution.

to those of Doublier and Wood [58], in which the storage modulus $G'$ was lower than the loss modulus $G''$ at low frequencies. As the frequency increased, because the increase rate of $G'$ was greater than that of $G''$, $G'$ took over $G''$ at a certain frequency, therefore, the $G'$ was greater than the $G''$ at higher frequencies. At low frequencies, where there is sufficient time for entanglements to make and break within the period of oscillation, the response is similar to that of non-entangled diluted solutions: $G'' \gg G'$, where $G' \propto \omega^2$ and $G'' \propto \omega$. In particular, $\eta^*$ is independent of $\omega$, giving a Newtonian plateau at low frequencies, analogous to that observed at low shear rates under unidirectional shear. At higher frequencies, as the rate of oscillation exceeds the timescale of molecular rearrangement, entanglement coupling becomes less distinguishable from "permanent" association of chains through gel "junction zones." The overall response approaches that of a gel, i.e., $G' > G''$, with little frequency dependence for either modulus and with $\eta^*$ decreasing steeply with increasing $\omega$ [57]. This mechanical spectrum at room temperature and normal frequency range falls into the "terminal zone" of the time-frequency master curve: the logarithmic plot of $G'$ and $G''$ against frequency [69].

However, partially hydrolyzed oat β-glucans exhibited some unusual viscoelastic behaviors as shown in Figure 3.31 [58]. The logarithmic plot of $G'$ and $G''$ against frequency of OG15 and OG60 (MW $\sim$ 300,000 and 70,000, respectively) showed unusual spectra when analyzed at the same temperature and frequency range as those of unhydrolyzed β-D-glucans, as shown in Figure 3.31 and Figure 3.32. At lower frequencies, the $G'$ of OG60 (2.75%, v/v) was higher than $G''$, typical of a gel structure. The $G''$ increased significantly faster

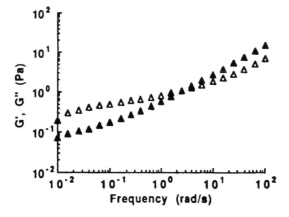

**Figure 3.31** Storage modulus ($G'$) and loss modulus ($G''$) as a function of angular frequency of acid-hydrolyzed oat gum (OG60) solutions. Concentration: 2.75% ($\Delta = G'$, ▲ $= G''$). Reprinted from Reference [58].

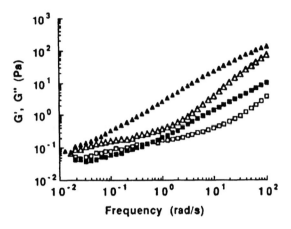

**Figure 3.32** Storage modulus ($G'$) and loss modulus ($G''$) as a function of angular frequency of acid-hydrolyzed oat gum (OG15) solutions. Concentrations: 2.75% ($\Delta = G'$, $\blacktriangle = G''$), 1.37% ($\square = G'$, $\blacksquare = G''$). Reprinted from Reference [58].

than $G'$, and as a result, $G''$ took over $G'$ at frequency 3 s$^{-1}$. The shape of this mechanical spectrum is similar to the "transition zone" in the master curve of $G'$ and $G''$ against frequency [58]. However, it is not possible to obtain the true transition zone under the reported experimental conditions. Therefore, the apparent transition zone may reflect some intermediate stage of the system, in which the response of $G'$ and $G''$ deviated from a normal, stable polymer system. Indeed, time dependence of the $G'$ and $G''$ of OG15 and OG60 solutions were observed (see the following two sections). A similar mechanical spectrum was observed for OG15 at 1.37% (v/v), while the taking over of $G''$ over $G'$ occurred at a lower frequency (0.7 s$^{-1}$); when the concentration is increased to 2.75%, the apparent transition zone is shifted to a lower frequency (Figure 3.32). A similar observation was reported by Gomez et al. [70] on barley $\beta$-D-glucan, that the appearance of the mechanical spectrum of $\beta$-D-glucan A (0.16%, Mw ~573,000) at room temperature was similar to that of OG15.

These observations are counter to the general rule of uncross-linked linear polymers. According to Ferry [69], the terminal zone of uncross-linked polymers progressively shifts to lower frequencies with increasing molecular weight. The sudden change from a terminal zone at a higher Mw to the plateau zone (gel-like structure), sometimes with an appearance similar to the transition zone for lower Mw $\beta$-D-glucan solutions, indicates that high and low Mw species of $\beta$-D-glucans have totally different rheological properties. It appeared that the degradation of $\beta$-D-glucan molecules enabled the formation of ordered conformation that is responsible for the unusual rheological behaviors of $\beta$-D-glucan solutions.

## 4.4. GELLING PROPERTIES OF β-D-GLUCANS

It has been noticed in the brewing industry that barley β-D-glucans form gels under storage conditions [3,4]. In our laboratory, we observed that commercial β-D-glucans (Megazyme Intl. MW ~200,000–300,000) and wheat β-D-glucans (Mw 300,000–600,000) at 2.0% (w/w) can form gels when stored at 4°C for a period of two to five days. The gel-forming ability of β-D-glucan samples of different molecular weights from oat, barley and wheat was examined at 2.0% (w/w) and 4°C. After 24 hr, wheat β-glucan solution formed a firm gel with a small but visible amount of water separated from the gel. The mechanical spectrum of the gel is shown in Figure 3.33 [15]. This gel starts to melt at 35°C, and the melting process is completed at 60°C (Figure 3.34). A similar gelling behavior was reported for a degraded barley β-D-glucan (BG165) but at a much higher concentration (6%, w/w) (Figure 3.35). The mechanical spectrum of freshly prepared BG165, 6% w/w is in the terminal zone of the master curve where the $G' < G''$ and both moduli are highly frequency dependent. However, when the same sample is stored for three days, it becomes a typical gel, and the mechanical spectrum falls in the plateau zone where $G' \gg G''$ and both moduli are frequency independent. The gel development rate of barley β-D-glucans varies with concentration and molecular weight. Bohm and Kulicke introduced

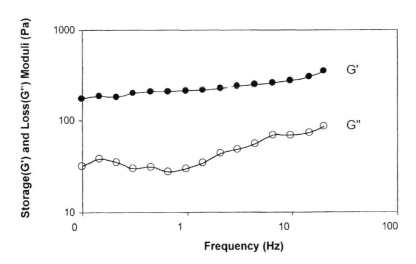

**Figure 3.33** Mechanical spectrum of 2.0% (w/w) wheat β-glucan gel at 4°C (Bohlin CVO Rheometer, Cone/plate 4°C). Reprinted from Reference [58] with permission. © 1995 Department of Agriculture and Agri-Food, Government of Canada.

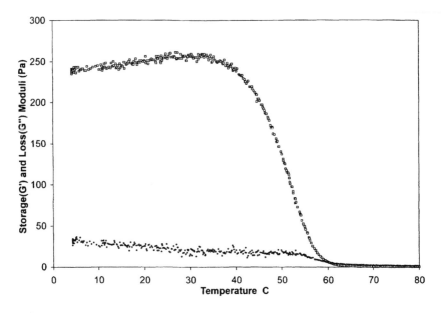

**Figure 3.34** Melting curve of 2.0% wheat β-glucan gel (w/w) at 1 Hz of frequency on a Bohlin CVO Rheometer (cone/plate 4°C) [28].

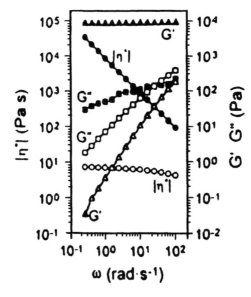

**Figure 3.35** Complex viscosity [η]. Storage modulus $G'$ and loss modulus $G''$ vs. angular frequency ω of a freshly prepared BG165, 6% (w/w) solution (open symbols) and the same sample three days later (solid symbols). Reprinted from Reference [37] with permission from Elsevier Science.

154

a new parameter, "elasticity increment" $I_E$, to measure the gelation rate:

$$I_E = \left(\frac{d\log G'}{dt}\right)_{max} \tag{3-7}$$

It indicates the maximum number of decades $G'$ increases per time unit. A high $I_E$ value reflects rapid gelation. The gelation rate is determined by a combination of the mobility of the polymer chains and the entanglement density in solutions: it declines with decreasing concentration and increasing molecular weight. For the same polymer, an increase in concentration will increase entanglement density, hence, the probability of interchain association, which is a basic requirement for three-dimensional network formation. At the same concentration, higher molecular weight β-D-glucans give lower $I_E$ numbers, i.e., slower rates of gel formation (Figure 3.36). It is believed that smaller molecules have a higher mobility, therefore accelerating the gelation process [56]; similar results and explanations were given to partially hydrolyzed oat β-D-glucans [58]. Bohm and Kulicke [56] also observed that polydispersity (Mw/Mn) played an important role in the gelation process. For example, the Mw of DG165 is higher than that of BG100 (165,000 vs. 100,000), but the $I_E$ of BG165 is higher than that of BG100. This seems to contradict the molecular mobility theory. However, the polydispersity of BG165 (1.9) is much higher than that of BG100 (1.3). The smaller molecules in this sample may form an entanglement nucleus. In order to verify the effect of polydispersity on gelation rate, Bohm and Kulicke [56] prepared three blends from β-D-glucan samples, BG50 and BG300, at different ratios. A strong influence of the polydispersity

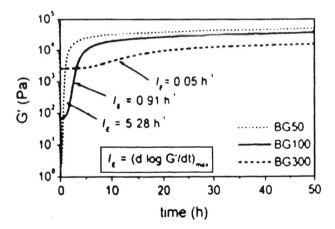

**Figure 3.36** Storage modulus $G'$ as a function of time for barley β-glucan samples of different molar mass in 10% (w/w) and the elasticity increment ($I_E$) values as a measure of gelation rate. Reprinted from Reference [37] with permission from Elsevier Science.

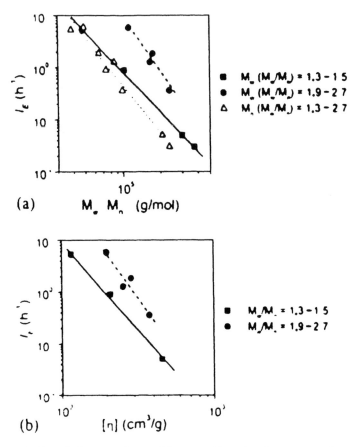

**Figure 3.37** Elasticity increment $I_E$ vs. weight-average Mw and number-average molar masses Mn (a) and intrinsic viscosities [η] (b) of several barley β-glucan samples in 10% (w/w) of narrow (1.3–1.5) and broad (1.9–2.7) polydispersities Mw/Mn. Reprinted from Reference [37] with permission from Elsevier Science.

was detected: broadly dispersed samples exhibited higher gelation rates than narrowly dispersed samples, as shown in Figure 3.37 [56]. A linear relationship was found between $I_E$ and log Mn:

$$I_E = 14.8 - 3.3 \log \text{Mn} \qquad (3\text{-}8)$$

Mn, which was measured by osmetrical method, emphasizes the low molecular weight portion of a polydispersed polymer and was believed to promote the gelation process [56].

Heat-prepared oat β-D-glucans and lichenan solutions also yielded gels similar to barley β-D-glucans but with different turbidity and gel melting

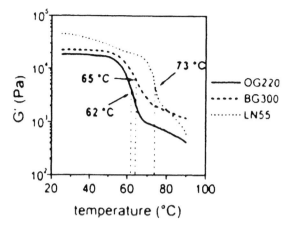

**Figure 3.38** Gel melting (step in storage modulus $G'$-curve) of oat β-glucan (OG220). Barley β-glucan (BG300) and lichenan (LN55) in 10% (w/w). Reprinted from Reference [37] with permission from Elsevier Science.

temperatures, as shown in Figure 3.38 [56]. At the same polymer concentration (10%, w/w), the melting temperature of β-D-glucans follows the order of lichenan (73°C), barley (65°C) and oat (62°C), which coincides with the order of tri- to tetrasaccharides ratio (22, 3 and 2, respectively) (Table 3.13). The gelation rate of the three β-D-glucans also differed dramatically ($I_E$ LN55 > 55 h$^{-1}$, BG50 = 5.3 h$^{-1}$ and OG40 = 0.86 h$^{-1}$), although they all have the same average molecular weight Mn (Mn = 30,000 for all three samples) [56]. The striking differences in gelation properties of the three β-D-glucans must have a great deal to do with their basic chemical structure. As we discussed in section 3, the main

TABLE 3.13. Comparison of Melting Temperature of
β-D-Glucan Gels and Gels from Commercial Gelling Agents.

| Sample Code | Concentration (% ww) | Melting Temperature (°C) |
|---|---|---|
| BG300 | 10 | 65 |
| BG165 | 10 | 65 |
| BG100 | 10 | 65 |
| OG220 | 10 | 62 |
| LN55 | 10 | 73 |
| Gelatin | 10 | 32 |
| $\iota$-Carrageenan | 4 | 40 |
| $\kappa$-Carrageenan | 4 | 72 |
| Agarose | 4 | 78 |

Reprinted from Reference [56] with permission from Elsevier Science.

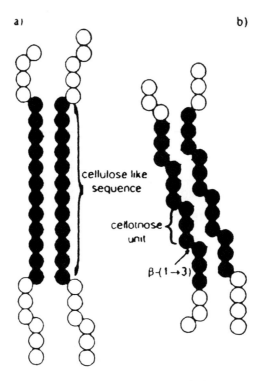

**Figure 3.39** Chain interactions in $(1\rightarrow3)(1\rightarrow4)$-β-glucan leading to gels. According to the common model, sequences of consecutive $(1\rightarrow4)$-linkages stick together (a). The authors propose association of consecutive cellotriose units linked by β-$(1\rightarrow3)$ bonds, which probably forms a helical structure (b). Reprinted from Reference [37] with permission from Elsevier Science.

structural feature of β-D-glucans is the trisaccharide unit interrupted mostly by the tetrasaccharide unit and less frequently by longer consecutive $(1\rightarrow4)$ linked β-D-glucosyl units (DP up to 14). Bohm and Kulicke assumed that the consecutive cellotriose units were responsible for the association. Lichenan has the highest proportion of the consecutive cellotriosyl unit, therefore, the highest tendency to form extended junction zones. The structure of oat β-D-glucan is much less regular so that the possibility of having consecutive cellotriosyl units is much lower, hence, the gelation rate [56]. An association model was proposed based on the above hypothesis; association of consecutive cellotriosyl units probably forms a helical structure, as shown in Figure 3.39 [56].

The relative gel strength (relative to BG50) expressed by $G'$ is shown in Table 3.14. Gels from β-D-glucans reveal syneresis that might be caused by phase separation into polymer-rich and polymer-weak regions. In the polymer-weak region, the mesh width is bigger and the water-holding capacity is poor, therefore, syneresis occurs [56].

TABLE 3.14. Comparison of Gel Strength ($G'$) and Relative Gel Strength (Relative to BG50) of Barley β-D-Glucan Gels and Four Gels Prepared from Commercial Gelling Agents.

| Sample Code | Concentration (% w/w) | $G'_p$ (Pa) | Rel. Gel Strength |
|---|---|---|---|
| BG50 | 6 | 6000 | 1·0 |
| BG100 | 6 | 7000 | 1·5 |
| BG165 | 6 | 8500 | 1·9 |
| BG300 | 6 | 7500 | 3·0 |
| BG100 | 4 | 1300 | 0·5 |
| BG100 | 8 | 18,000 | 3·6 |
| BG100 | 10 | 42,000 | 5·5 |
| Gelatin | 6 | 400 | 0·6 |
| $_l$-Carrageenan | 6 | 1000 | 0·6 |
| $_k$-Carrageenan | 6 | 70,000 | 7·4 |
| Agarose | 6 | 120,000 | 16 |

Reprinted from Reference [56] with permission from Elsevier Science.

## 4.5. DISCUSSION OF STRUCTURE AND PROPERTY RELATIONSHIPS OF β-D-GLUCANS

In summary, β-D-glucan is a linear polymer that exhibits a wide range of rheological properties, including viscoelastic fluid, weak gel and real gel. These rheological properties and other physical properties (e.g., solubility) depend on two important structural features: molecular weight and structural regularity (expressed by tri/tetra ratio). The higher the tri/tetra ratio, the more regular the structure, because a higher tri/tetra ratio gives a higher possibility of having more consecutive trisaccharide units that are believed to be essential for the formation of an ordered structure. In order to verify the work of Doublier and Wood [58], we re-examined the rheological properties of OG15 and OG60. Freshly prepared samples (3%, w/w) give a mechanical spectra of typical viscoelastic fluid (Figure 3.40). However, monitoring of $G'$ and $G''$ against time at 0.1 Hz revealed that the $G'$ steadily increased with time, while the $G''$ remained relatively unchanged. A mechanical spectrum similar to that of Doublier and Wood's apparent transition zone was obtained after 18 hr (Figure 3.41). Allowing the gel to further develop for another 10 hr, a firm gel structure that had a sharp melting point was evident (Figure 3.42).

The relationship between the structural regularity and gelation rate may also be related to the correlation between the amount of ammonium sulphate necessary for the precipitation of barley β-D-glucans from aqueous solution with the regularity of the polymer chain. The fractions precipitated at a low concentration of ammonium sulphate (fairly soluble) contained more cellotriosyl units

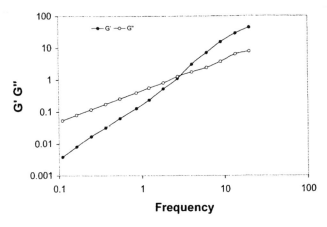

**Figure 3.40** Mechanical spectrum of freshly prepared 3.0% (w/w) β-D-glucan OG60 (MW 70,000) [72].

than the readily soluble fractions [56,71]. The solubility of freeze-dried β-D-glucans discussed in section 4.3 also follows the same order and is summarized in Table 3.15 [15]. Three relationships can be drawn:

(1) High MW β-D-glucans (above 800,000) exhibit viscoelastic fluid flow behavior in aqueous systems, while the rheological properties of the low molecular weight species (40,000–600,000) are subject to change with time.

(2) Freshly prepared low MW β-D-glucan solutions are similar to high Mw species, as both exhibit viscoelastic fluid characteristics; however, as time

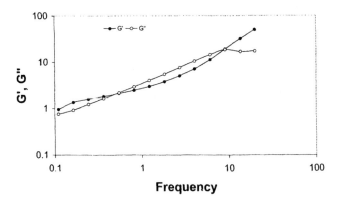

**Figure 3.41** Mechanical spectrum of 3.0% (w/w) β-D-glucan OG60 (MW 70,000) after 12 hr of preparation [72].

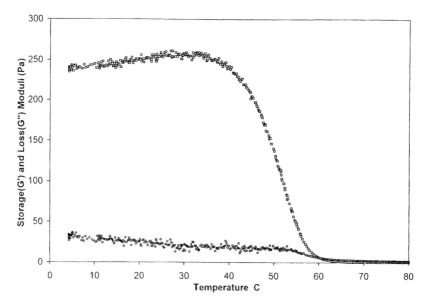

**Figure 3.42** Melting curve of 3.0% (w/w) β-D-glucan OG60 (MW 70,000) gel after 24 hr of preparation [72].

passes, the low Mw β-D-glucans tend to form ordered structures in solutions, which essentially leads to formation of gels. The determinations of rheological properties at any time interval only reflect the properties at that particular moment.

(3) The gelation rate is inversely proportional to Mn (presumably, there should also be a minimum molecular weight in order to form a gel).

(4) At the same molecular weight range, the tri/tetra ratio of β-D-glucans affects the properties related to the formation of ordered structures, such as gelation

**TABLE 3.15.** Comparison of Structural Features and Physical Properties of Cereal β-Glucans [15].

| β-D-Glucan Source | MW ×10³ | Ratio of Tri/Tetra | Freeze-Dried Sample Solubility | Aggregation/Gel Formation Ability |
|---|---|---|---|---|
| Oat (high Mw) | 976 | 2.2 | Difficult | Low |
| Oat (low Mw) | 263 | 2.2 | Difficult | Medium |
| Barley (high Mw) | 767 | 3.0 | Difficult | Low |
| Barley (medium) | 290 | 3.0 | 70°C, 1–2 hr | Medium high |
| Barley (low Mw) | 72–213 | 3.0 | 90°C, several hr | Medium high |
| Wheat | 300 | 4.5 | Insoluble | High |

rate and insolubility in water and salt solutions. The higher the tri/tetra ratio, the faster the gelation rate and the more difficult it is to dissolve them in solvents.

## 5. APPLICATIONS

β-D-glucans have not been applied in foods as ingredients or stabilizers. But, as an active component in oat products, such as whole oats, oat bran, oatmeal, etc., oat β-D-glucans have been used extensively in breakfast cereals and snack foods. Recently, purified oat β-D-glucan solution was used in the cosmetic industry as a moisturizing agent and a thickener. Extensive research is needed in this area to further expand the applications of β-D-glucans for food and non-food uses. For more information on the physiological effects of cereal β-D-glucans, readers are referred to a chapter by Wood and Beer [8].

## 6. REFERENCES

1. Fulcher R. G., Miller S. S. Structure of oat bran and distribution of dietary fiber components. *Oat Bran*. Wood P. J., ed. St. Paul, MN: Am. Assoc. Cereal Chem.; 1993. pp. 1–24.

2. Autio K. *Functional Aspects of Cereal Cell Wall Polysaccharides*. In *Carbohydrates in Food*. A-C Eliason, ed. New York, Basel, Hong Kong: Marcel Dekker Inc.; 1996. pp. 227–264.

3. Bamforth C. W. Barley β-glucans, their role in malting and brewing. *Brew. Dig.* 1982; 57:22.

4. Bamforth C. W. Biochemical approaches to beer quality. *J. Inst. Brew.* 1985; 91:154–160.

5. Rotter B. A., Marquaedt R. R., Guenter W., Biliaderis C. G., Newman W. *In vitro* viscosity measurements of barley extracts as predictors of growth responses in chicks fed barley-based diets supplemented with fungal enzyme preparation. *Can. J. Anim. Sci.* 1989; 69:431.

6. Thacker P. A., Campbell G. L., GrootWassink J. W. D. The effect of sodium bentonite on the performance of pigs fed barley-based diets supplemented with beta-glycanase. *Nutr. Rep. Int.* 1989; 40:613.

7. Thacker P. A., Campbell G. L., Groot Wassink J. W. D. The effect of beta-glucanase supplementation on the performance of pigs fed hulled barley. *Nutr. Trp. Int.* 1988; 38:91.

8. Wood P. J., Beer M. Oat β-glucan. In *Functional Foods*. Mazza G., ed. Lancaster, PA; Technomic Publishing Co., Inc.; 1998. pp. 1–37.

9. Wood P. J., Braaten J. T., Scott F. W., Riedel D., Poste L. M. Comparisons of viscous properties of oat and guar gum and the effects of these and oat bran on glycemic index. *J. Agric. Food Chem.* 1990; 38:753–7.

10. Braaten J. T., Wood P. J., Scott F. W., Riedel K. D., Poste L. M., Collins M. W. Oat gum, a soluble fiber which lowers glucose and insulin in normal individuals after an oral glucose load: comparison with guar gum. *Am. J. Clin. Nutr.* 1991; 53:1425–1430.

11. Roemer T., Paravicini G., Payton M. A., Bussey H. Characterization of the yeast $(1 \rightarrow 6)$-beta-glucan biosynthetic components, Kre6P and Skn1P, and genetic interactions between the Pkci pathway and extracellular-matrix assembly. *J. Cell Biology* 1994; 127(2):567–79.

12. Cui W., Wood P., Weisz J., Beer M. U. Non-starch polysaccharides from pre-processed wheat bran: extraction and physicochemical characterizations. *Cereal Chem.* 1999; 76:129–33.

13. Wood P. J. Physicochemical characteristics and physiological properties of oat $(1\rightarrow3)(1\rightarrow4)$-β-glucan. *Oat Bran.* Wood P. J., ed. St. Paul, MN: Amer. Assoc. Cereal Chem.; 1993. pp. 83–112.

14. Wood P. J., Weisz J., Blackwell B. A. Structural studies of $(1\rightarrow3)(1\rightarrow4)$-β-D-glucans by $^{13}$C-NMR and by rapid analysis of cellulose-like regions using high-performance anion-exchange chromatography of oligosaccharides released by lichenase. *Cereal. Chem.* 1994; 71:301–307.

15. Cui W., Wood P. J. Relationships between structural features, molecular weight and rheological properties of cereal β-D-glucan. *The 4th International Conference on Hydrocolloids*, Oct 4–10, 1998, Osaka, Japan.

16. Beer M. U., Arrigoni E., Amado R. Extraction of oat gum from oat bran—effects of process on yield, molecular-weight distribution, viscosity and $(1\rightarrow3)(1\rightarrow4)$-beta-D-glucan content of the gum. *Cereal Chem.* 1996; 73:58–62.

17. Wood P. J., Weisz J., Blackwell B. A. Molecular characterization of cereal beta-D-glucans—structural analysis of oat beta D glucan and rapid structural evaluation of beta-D glucans from different sources by high-performance liquid-chromatography of oligosaccharides released by lichenase. *Cereal Chem.* 1991; 68:31–9.

18. McCleary B. V. Purification of $(1\rightarrow3),(1\rightarrow4)$-beta-D-glucan from barley flour. *Methods in Enzymology* 1988; 160:511–4.

19. Aman P., Graham H., Tilly A. C. Content and solubility of mixed-linked $(1\rightarrow3)$, $(1\rightarrow4)$-beta-D-glucan in barley and oats during kernel development and storage. *J. Cereal Sci.* 1989; 10:45–50.

20. Anderson M. A., Cook J. A., Stone B. A. Enzymatic determination of $(1\rightarrow3)(1\rightarrow4)$-β-glucans in barley grain and other cereals. *J. Inst. Brew.* 1978; 84:233–239.

21. Beer M. U., Wood P. J., Weisz J. Molecular-weight distribution and $(1\rightarrow3)(1\rightarrow4)$-beta-D-glucan content of consecutive extracts of various oat and barley cultivars. *Cereal Chem.* 1997; 74:476–80.

22. Carr J. M., Glatter S., Jeraci J. L., Lewis B. A. Enzymatic determination of beta-glucan in cereal-based food-products. *Cereal Chem.* 1990; 67:226–9.

23. Jorgensen K. G. Quantification of high molecular-weight $(1\rightarrow3)(1\rightarrow4)$-beta-D-glucan using calcofluor complex-formation and flow-injection analysis. 1. Analytical principle and its standardization. *Carlsberg Research Communications* 1988; 53(5):277–85.

24. Ahluwahlia B., Ellis E. E. A rapid and simple method for the determination of starch and β-glucan in barley and malt. *J. Inst. Brew.* 1984; 90:254–259.

25. Varum K. M., Smidsrod O. Partial chemical and physical characterisation of $(1\rightarrow3)(1\rightarrow4)$-β-D-glucans from oat (*Avena sariva* L.) aleurone. *Carbohydr. Polym.* 1988; 9:103–117.

26. Woodward J. R., Phillips D. R., Fincher G. B. Water-soluble $(1\rightarrow3),(1\rightarrow4)$-β-glucans from barley (*Hordeum vulgare*) endosperm. IV. Comparison of 40°C and 65°C soluble fractions. *Carbohydr. Polym.* 1988; 8:85–97.

27. Woodward J. R., Phillips D. R., Fincher G. B. Water-soluble $(1\rightarrow3),(1\rightarrow4)$-β-D-glucans from barley (*Hordeum vulgare*) endosperm. I. Physicochemical properties. *Carbohydr. Polym.* 1983; 3:143.

28. Cui W., Wood P. J., Weisz J., Mullin J. Unique gelling properties of non-starch polysaccharides from pre-processed wheat bran. In *Gums and Stabilizers for the Food Industry 9*. P. A. Williams and G. O. Phillips, eds. The Royal Society of Chemistry; 1998. pp. 34–42.

29. Beer M. U., Wood P. J., Weisz J., Fillion N. Effect of cooking and storage on the amount and molecular-weight of $(1\rightarrow3)(1\rightarrow4)$-beta-D-glucan extracted from oat products by an *in-vitro* digestion system. *Cereal Chem.* 1997; 74:705–9.

30. Wood P. J., Fulcher R. G. Interaction of some dyes with cereal β-glucans. *Cereal Chem.* 1978; 55:952–966.

31. Beresford G., Stone B. A. (1→3),(1→4)-beta-D-glucan content of triticum grains. *J. Cereal Sci.* 1983; 1:111–4.

32. Westerlund E., Andersson R., Aman P. Isolation and chemical characterization of water-soluble mixed-linked β-glucans and arainoxylans in oat milling fractions. *Carbohydr. Polym.* 1993; 20:115–23.

33. Wood P. J. Oat β-glucan: structure, location and properties. In *Oats: Chemistry and Technology.* Webster F. H., ed. St. Paul, MN: Am. Assoc. Cereal Chem.; 1986. pp. 121–152.

34. Wood P. J., Weisz J., Fedec P., Burrows V. D. Large-scale preparation and properties of oat fractions enriched in (1→3)(1→4)-beta-D-glucan. *Cereal Chem.* 1989; 66:97–103.

35. Wood P. J., Weisz J., Fedec P. Potential for beta-glucan enrichment in brans derived from oat (*Avena sativa* L.) cultivars of different (1→3),(1→4)-beta-D-glucan concentrations. *Cereal Chem.* 1991; 68:48–51.

36. Preece I. A., Hobkirk R. Non-starchy polysaccharides of cereal grains III. Higher molecular gums of common cereals. *J. Inst. Brew.* 1953; 59:385–92.

37. Bohm N., Kulicke W. M. Rheological studies of barley (1→3)(1→4)-β-glucan in concentrated solution: investigation of the viscoelastic flow behaviour in the sol-state. *Carbohydr. Res.* 1999; 315:293–301.

38. Malkki Y., Autio K., Hanninen O., Myllymaki O., Pelkonen K., Suortti T., Torronen R. Oat bran concentrates—physical-properties of beta-glucan and hypocholesterolemic effects in rats. *Cereal Chem.* 1992; 69:647–53.

39. Autio K., Myllymaki O., Suortti T., Saastamoinen M., Poutanen K. Physical-properties of (1→3),(1→4)-beta-D-glucan preparates isolated from Finnish oat varieties. *Food Hydrocolloids* 1992; 5:513–22.

40. Dais P., Perlin A. S. High-field, $^{13}$C-N.M.R. spectroscopy of β-D-glucan, amylopectin, and glycogen. *Carbohydr. Res.* 1982; 100:103–16.

41. Cui W., Wood P. J., Blackwell B., Nikiforuk J. Physicochemical properties and structural characterization by 2 dimensional NMR spectroscopy of wheat β-D-glucan-comparison with other cereal β-D-glucans. *Carbohydr. Polym.* 2000; 41:249–258.

42. Wood P. J. Specificity in the interaction of direct dyes with polysaccharides. *Carbohydr. Res.* 1980a; 85:271–87.

43. Wood P. J., Fulcher R. G., Stone B. A. Studies on the specificity of interaction of cereal cell-wall components with Congo Red and Calcofluor—specific detection and histochemistry of (1→3),(1→4),-beta-D-glucan. *Journal of Cereal Science* 1983; 1:95–110.

44. Jorgensen K. G., Jensen S. A. S., Hartley P., Munck L. The analysis of β-glucan and β-glucanase in malting and brewing using Calcofluor and Congo Red. *EBC Congress*, Helsiuri; 1985.

45. Wood P. J. Factors affecting precipitation and spectral changes associated with complex formation between dyes and β-glucan. *Carbohydr. Res.* 1982; 102:283–93.

46. Wood P. J., Anderson J. W., Braaten J. T., Cave N. A., Scott F. W., Vachon C. Physiological-effects of beta-D-glucan rich fractions from oats. *Cereal Foods World* 1988; 33:678.

47. McCleary B. V., Nurthen E. Measurement of (1→3)(1→4)-beta-D-glucan in malt, wort and beer. *Journal of the Institute of Brewing* 1986; 92:168–73.

48. McCleary B. V., Glennie-Holmes M. Enzymic quantification of (1→3),(1→4)-β-D-glucan in barley and malt. *J. Inst Brew* 1985; 91:285–95.

49. McCleary B. V., Codd R. Measurement of $(1\rightarrow3),(1\rightarrow4)$-beta-D-glucan in barley and oats— a streamlined enzymatic procedure. *J. Sci. Food and Agric.* 1991; 55:303–12.

50. Varum K. M., Martinsen A., Smidsrod O. Fractionation and viscometric characterisation of a $(1\rightarrow3)(1\rightarrow4)$-β-glucan from oat, and universal calibration of a high-performance size exclusion chromatographic system by the use of fractionated β-glucans, alginates and pullulans. *Food Hydrocolloids* 1991; 5:363–373.

51. Forrest I. S., Wainwright T. The mode of binding of β-glucans and pentosans in barley endosperm cell walls. *J. Inst. Brew.* 1977; 139:535–545.

52. Varum K. M., Smidsrod O., Brant D. Light scattering reveals micelle-like aggregation in the $(1\rightarrow3)(1\rightarrow4)$-β-D-glucans from the oat aleurone. *Food Hydrocolloids.* 1992; 4:497–511.

53. Suortti T. Size-exclusion chromatographic determination of beta-glucan with postcolumn reaction detection. *J. Chrom.* 1993; 632:105–10.

54. Bhatty R. S. Laboratory and pilot plant extraction and purification of β-glucans from hull-less barley and oat brans. *J. Cereal Sci.* 1995; 22:163–170.

55. Zhang D., Moore W. R. Influence of heat pretreatments of oat grain on the viscosity of flour slurries. *J. Sci. Food Agric.* 1997; 74:125–31.

56. Bohm N., Kulicke W. M. Rheological studies of barley $(1\rightarrow3)(1\rightarrow4)$-β-glucan in concentrated solution: mechanistic and kinetic investigation of the gel formation. *Carbohydr. Res.* 1999; 315:302–11.

57. Morris E. R. Polysaccharide solution properties: origin, rheological characterization and implications for food systems. In *Frontiers in Carbohydrate Research, 1: Food Application.* Millane R. P., BeMiller J. N., Chandrasekaran R., eds. New York: Elsevier; 1989. pp. 132–163.

58. Doublier J. L., Wood P. J. Rheological properties of aqueous-solutions of $(1\rightarrow3)(1\rightarrow4)$-beta-D-glucan from oats (*Avena sativa* L.). *Cereal Chem.* 1995; 72:335–40.

59. Grimm A., Kruger E., Burchard W. Solution properties of beta-D-(1,3)(1,4)-glucan isolated from beer. *Carbohydr. Polym.* 1995; 27:205–14.

60. Gomez C., Navarro A., Manzanares P., Horta A., Carbonell J. V. Physical and structural-properties of barley $(1\rightarrow3),(1\rightarrow4)$-beta-D-glucan. 1. Determination of molecular-weight and macromolecular radius by light-scattering. *Carbohydr. Polym.* 1997; 32:7–15.

61. Gomez C., Navarro A., Manzanares P., Horta A., Carbonell J. V. Physical and structural-properties of barley $(1\rightarrow3),(1\rightarrow4)$-beta-D-glucan. 2. Viscosity, chain stiffness and macromolecular dimensions. *Carbohydr. Polym.* 1997; 32:17–22.

62. Tvaroska I., Deslandes Y., Marchessault R. H., Ogawa K. Crystalline conformation and structure of lichenan and barley beta-glucan. *Can. J. Chem.* 1983; 61:1608–16.

63. Autio K. Rheological properties of solutions of oat β-glucans. In *Gums and Stabilisers for the Food Industry 4.* Phillips G. O., Wedlock D. J., Williams P. A., eds. Oxford: IRL Press; 1988. pp. 483–8.

64. Autio K., Myllymaki Y., Malkki Y. Flow properties of solutions of oat β-glucans. *J. Food Sci.* 1988; 52:1364–1366.

65. Wikstrom K., Lindahl L., Andersson R., Westerlund E. Rheological studies of water-soluble $(1\rightarrow3),(1\rightarrow4)$-β-D-glucans from milling fractions of oat. *J. Food Sci.* 1994; 59:1077–1080.

66. Cox W. P., Merz E. H. Correlation of dynamic and flow viscosities. *J. Polym. Sci.* 1958; 28:619–22.

67. Navarini L., Cesaro A., Ross-Murphy S. B. Exopolysaccharides from *Rhizobium muliloti* YE-2 grown and different osmolarity conditions: viscoelastic properies. *Carbohydr. Res.* 1992; 223:227–34.

68. Cui W., Eskin N. A. M., Biliaderis C. G. Water-soluble yellow mustard (*Sinapis alba* L.) polysaccharides—partial characterization, molecular-size distribution and rheological properties. *Carbohydr. Polym.* 1993; 20:215–25.

69. Ferry J. D. *Viscoelastic Properties of Polymers.* New York: John Wiley & Sons, Inc.; 1960.

70. Gomez C., Navarro A., Garnier C., Horta A., Carbonell J. V. Physical and structural-properties of barley $(1\rightarrow3),(1\rightarrow4)$-beta-D-glucan III—formation of aggregates analyzed through its viscoelastic and flow behavior. *Carbohydr. Polym.* 1997; 34:141–8.

71. Izawa M., Kano Y., Koshino S. Relationship between structure and solubility of $(1\rightarrow3)(1\rightarrow4)$-β-D-glucan from barley. *J. Am. Soc. Brew. Chem.* 1993; 51:123–7.

72. Cui W., Wang Q., Wood P. J. Time dependence of rheological properties of partially hydrolyzed oat gum. 1999, unpublished data.

# CHAPTER 4

# Cereal Non-Starch Polysaccharides II: Pentosans/Arabinoxylans

## 1. INTRODUCTION

"**W**E know now that they (hemicelluloses) possess valuable properties, resembling the commercial exudate gums. It is surprising that more effort has not been directed toward development of processes that allow the production of hemicellulose on a commercial scale" [1, p. 413]. These comments were made by Professor Roy L. Whistler 15 years ago at the international conference on "New Approaches to Research on Cereal Carbohydrates" held at the Carlsberg Research Center, Copenhagen, Denmark, June 24–29, 1984 [1]. Hemicelluloses and pentosans are exchangable terms used to designate non-starch polysaccharides (NSP) from cereals. Arabinoxylan has been used frequently because most of the pentosans are composed of a 1,4-linked-β-D-xylose backbone chain to which arabinose side chains are attached at the 2,3 positions. Pentosans/arabinoxylans are the primary construction materials of the cell walls of some cereals, such as wheat and rye. Whereas in oat and barley, $(1 \rightarrow 3)(1 \rightarrow 4)$ mixed linked-β-D-glucans (described in Chapter 3) are the major building components of the endosperm cell walls. These cell wall materials serve the growing plant as a structural component for maintaining tissue integrity, as a media for transporting water and other low molecular weight solutes and nutrients and as a barrier against microbe and insect penetration.

TABLE 4.1. **The Contents of Water-Soluble and Total Pentosans in Cereals.**

| Cereal Source | Water-Soluble Pentosans % | Total Pentosans % |
|---|---|---|
| Wheat | 0.3–0.8 | 4–6% |
| Rye | 0.74 | 7–8% |
| Barley | 0.4 | 4–8% |
| Oat | 0.2 | 2.2–4.1% |

Adapted from Reference [2,3].

The contents of the water-soluble and total pentosans in wheat are 0.3–0.8% and 4–6%, respectively [2], while their corresponding levels in rye are ~0.7% and 7–8% [3]. The content of pentosan in barley ranges from 4–8%, of which up to 75% is found in the husk and is insoluble in water [4,5]. The total and water-soluble pentosans in oat are the lowest, ranging from 2.2–4.1% and ~0.2%, respectively [2]. These results are summarized in Table 4.1.

Cereal pentosans are considered anti-nutrients in the feed industry because of their high viscosity in aqueous mediums, that interferes with the digestion and utilization of nutrients. In the food industry, pentosans are found to be closely related to the quality of baking products. For example, removal of water-soluble pentosans from wheat flour reduced the loaf volume of the resultant bread; in contrast, the addition of 2% water-soluble endosperm pentosans increased the loaf volume by 30–45%, and it also improved cell uniformity, crumb characteristics and elasticity [6]. High viscosity of pentosan solutions is considered extremely important for the quality of rye bread; it replaces the cell forming function of wheat gluten in rye dough [7]. Pentosan also stabilizes foams in beer, although it causes an undesirable effect on beer filtration and haze formation [8]. All of these functional properties, either positive or negative, are primarily determined by their structural features, including monosaccharide composition, linkage and substitution patterns and molecular weight. Much work has been carried out on wheat arabinoxylans; in contrast, little research has been done on arabinoxylans from barley, oat, rye and maize. This chapter reviews the most recent developments concerning the extraction processes, structural characteristics, physicochemical and functional properties of cereal arabinoxylans and their potential applications as stabilizers in the food industry.

## 2. EXTRACTION AND FRACTIONATION OF PENTOSANS

According to solubility in water, cereal pentosans are classified as water-soluble and water-insoluble (WSP and WIP, respectively). The water-soluble wheat pentosans are present mainly in flour (endosperm cell walls), while the

TABLE 4.2. **Content of Water-Soluble and Total Pentosans in Wheat Classes and Varieties.**

|  |  | Pentosan (%) | |
| --- | --- | --- | --- |
| No. | Cultivar | Water-Soluble | Total |
| Hard red and white |  |  |  |
| 1 | Batum | 0.683 | 5.396 |
| 2 | Hatton | 0.709 | 5.742 |
| 3 | Wanser | 0.648 | 5.773 |
| 4 | McKay | 0.559 | 5.651 |
| 5 | Wampum | 0.533 | 6.059 |
| 6 | Spillman | 0.633 | 5.762 |
| 7 | Burt | 0.568 | 4.839 |
| Soft white |  |  |  |
| 8 | Daws | 0.755 | 5.565 |
| 9 | Yamhill | 0.475 | 5.303 |
| 10 | Stephens | 0.499 | 5.51 |
| 11 | Nugaine | 0.549 | 5.671 |
| 12 | Dirkwin | 0.535 | 5.536 |
| 13 | Owens | 0.375 | 5.112 |
| 14 | WA7074 | 0.481 | 5.185 |
| White club |  |  |  |
| 15 | Tyee | 0.386 | 4.862 |
| 16 | Paha | 0.306 | 4.066 |
| 17 | Crew | 0.411 | 4.772 |
| 18 | Moro | 0.408 | 4.579 |

Reprinted from Reference [9].

water-insolubles can be found in bran and flour fractions. Variations of water-soluble and total pentosans in different wheat varieties are shown in Table 4.2 [9]. The distribution of water-soluble and insoluble pentosans in wheat and its milling products is summarized in Table 4.3 [10].

## 2.1. EXTRACTION OF WATER-SOLUBLE PENTOSANS (WSP)

### 2.1.1. Extraction from Wheat Flour

The contents of water-soluble and total pentosans in wheat flour are 0.5–0.7% and 1.3– 2.0%, respectively, and the level changes with classes and varieties, as shown in Table 4.4 [11]. The water-soluble fraction is extractable with cold water. For example, pentosans are extracted from wheat flour by mixing flour with water (1:3) and blending for 5 min. After centrifugation, the supernatant

TABLE 4.3. **Water-Soluble and Total Pentosan Contents in Milling Fractions from a Hard Red Winter Wheat.**

| | | Pentosan | |
| --- | --- | --- | --- |
| Mill Fraction[a] | | Soluble (%) | Total (%) |
| Whole Wheat | | 0.68 | 6.71 |
| Flour | 1,2 MDS | 0.53 | 1.13 |
| | SIZ,3 MDS | 0.56 | 1.15 |
| | 1,2,3 BRK | 0.62 | 1.29 |
| | 4 BRK,1,2 LG, 4,5, MDS | 0.59 | 1.59 |
| Bran | Coarsebran | 0.97 | 21.29 |
| | Finebran | 1.15 | 17.37 |

[a]Abbreviations: MDS = middlings, SIZ = sizings, BRK = break flour, LG = low-grade flour.
Reprinted from Reference [10].

is heated at 90–95°C for 30 min to deactivate endogenous enzymes. The deactivated enzymes are subsequently removed by absorption with celite and/or vega clay. The extraction procedure is demonstrated in Figure 4.1, in which the incubation with salivary α-amylase and the dialysis procedure were repeated

TABLE 4.4. **The Content of Total and Water-Soluble Pentosans in Different Wheat Flours.**

| Wheat Class and Variety | Water-Soluble Pentosans (%, db) | Total Pentosans (%, db) |
| --- | --- | --- |
| Canada prairie spring | | |
| HY 355 | 0.55 | 1.83 |
| HY 320 | 0.63 | 1.96 |
| Oslo | 0.50 | 1.60 |
| Canada utility | | |
| Glenlea | 0.55 | 1.67 |
| Canada western soft white spring | | |
| Fielder | 0.60 | 1.86 |
| Canada western red winter | | |
| Norstar | 0.54 | 1.37 |
| Unregistered (hard red spring) | | |
| Marshall | 0.62 | 1.76 |
| Canada western red spring | | |
| Katepwa | 0.68 | 2.06 |

Reprinted from Reference [11].

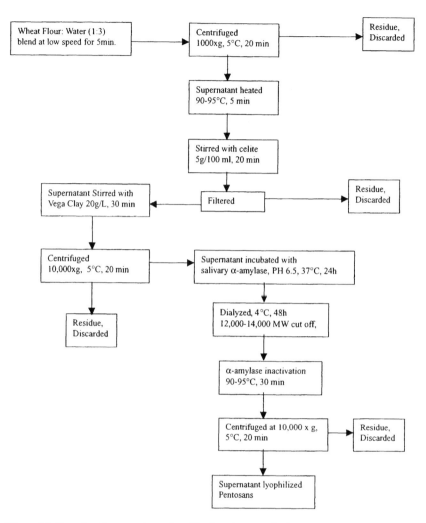

**Figure 4.1** Procedure for extraction and purification of water-soluble pentosans from wheat flour. Reprinted from Reference [12] with permission from the *Journal of Cereal Science:* © 1990 Academic Press Ltd.

three times to ensure complete elimination of small starch molecules [12]. The final product contained 6% proteins.

An alternative method for extracting water-soluble pentosans from wheat flour is to mix the flour with water to make dough. The dough is further treated with 0.1 M NaCl solution to form slurry. The slurry is centrifuged to produce a supernatant (the water-soluble fraction) and light and heavy tails. From the light and heavy tailings, respectively, Hoffmann [13] extracted pentosans with warm water to give a total yield of 1.1% of the flour, which includes 0.7% light pentosans and 0.4% heavy pentosans. The extraction procedures are shown in Figure 4.2, which also involved α-amylase digestion and pronase treatments to remove starches and proteins.

The extracted WSP is purified according to procedures described in Figure 4.3. The purified product is an arabinoxylan accounting for 0.5% of the

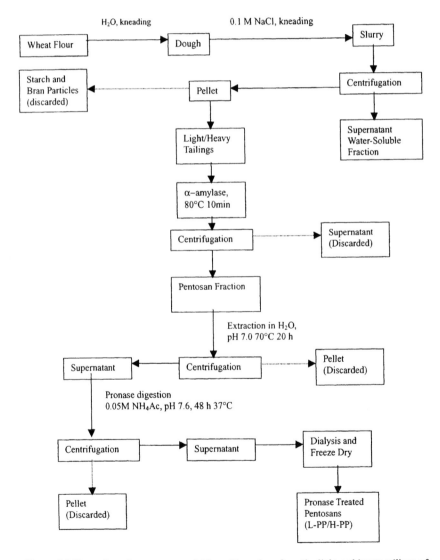

**Figure 4.2** Extraction of warm water-soluble arabinoxylans from the light and heavy tailings of wheat flour dough [13].

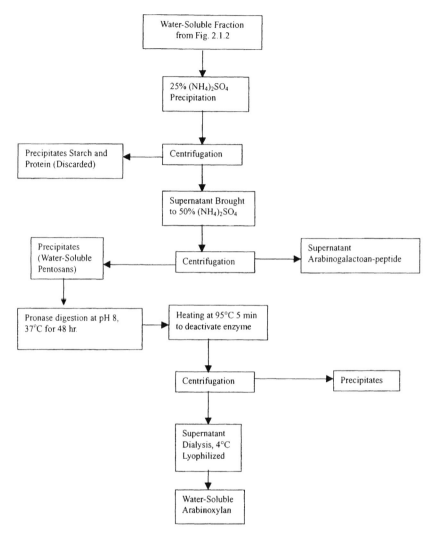

**Figure 4.3** Purification of water-soluble arabinoxylans from wheat flour dough [13].

wheat flour, which contained only trace amounts of protein contaminants (0.6% protein) [13].

## 2.1.2. Extraction of Water-Soluble Pentosans from Oat, Barley and Rye

Oat and barley arabinoxylans are usually co-extracted with β-D-glucan in water. The extraction process is similar to the procedure for β-D-glucans, as described in Chapter 3. The co-extracted non-starch polysaccharide mixture, as described by Wood et al. [14] and Westerlund et al. [15], is separated into arabinoxylans and $\beta$-D-glucan fractions by the $(NH_4)_2SO_4$ precipitation method [16,17]. Although this method was developed for the purification of cereal $\beta$-D-glucans, the arabinoxylans can be recovered from the supernatant of the $(NH_4)_2SO_4$ precipitation process.

Non-starch polysaccharides from barley can be sequentially extracted with water [18]. Barley grist (500 g) is extracted twice with 40°C water (1.5 L) for 30 min. The extracts are separated from the insoluble residue by centrifugation (5000 g, 20 min). The supernatants are combined and brought to 95°C for 5 min, then cooled to room temperature. The denatured protein during the heating process is removed by filtration through celite (20 g/L) and is further treated with Vega clay (Pembina Mountain Clay, Winnipeg, Manitoba) to remove the residue protein. Porcine pancreatic $\alpha$-amylase was incubated for 24 hr (pH 6.5, 35°C) to break down any starch in the extract and then underwent dialysis (Mw cutoff of 12,000–14,000). Purified extract is finally freeze-dried to give WE40 (1.4% of barley grist). The insoluble residue of the barley grist is suspended in 1 L water containing thermally stable $\alpha$-amylase (4500 U, Megazyme) and then heated at 65°C for 90 min. The extraction is repeated once, and the extracts are centrifuged and purified following the same procedure used for WE40. The purified and freeze-dried product is WE65, which accounts for 1.3% of the original barley grist [18]. The purified extracts, WE40 and WE65, respectively, are fractionated by a grade ammonium sulphate precipitation technique [19] to produce arabinoxylan-rich fractions (15% of WE40; 3% of WE65) and $\beta$-D-glucan-rich fractions (85% of WE40; 93% of WE65). Barley arabinoxylans are most soluble at 40°C.

The extraction and purification procedures used to obtain arabinoxylans from rye are similar to those for oat and barley. Ground rye meal is refluxed with 90% ethanol to remove low molecular weight compounds and to inactivate endogenous enzymes. The insoluble residue is isolated and washed with the same aqueous ethanol solvent. The residue is then extracted with water at 40°C for 90 min. In the middle of this treatment, the slurry is homogenized for 2 min to improve extractability. About 500 mL of water extract (A) is precipitated with 200 g $(NH_4)_2SO_4$. The precipitate (B) contained lesser amounts of glucose and arabinose. The supernatant fraction is further purified by dissolution in water and precipitation with 67% aqueous ethanol to give a crude arabinoxylan (C). Monosaccharide compositions of these three fractions are shown in Table 4.5 [20].

## 2.2. EXTRACTION OF WATER-INSOLUBLE PENTOSANS

### 2.2.1. Extraction of Water-Insoluble Pentosans from Wheat Flour

The preparation of water-insoluble pentosans from wheat flour requires two steps: (a) isolation of cell wall material from wheat flour and (b) use of alkaline

**TABLE 4.5.** **Monosaccharide Composition of Three Fractions during the Extraction and Fractionation Process of Rye Arabinoxylans.**

| Monosaccharide | Water-Extract (A) % | $(NH_4)_2SO_4$ Precipitate (B) % | Crude Arabinoxylan (C) % |
|---|---|---|---|
| Arabinose | 20.3 | 26.2 | 23.7 |
| Xylose | 33.3 | 54.1 | 48.3 |
| Glucose | 46.4 | 19.7 | 4.2 |

Adapted from Reference [20] with permission from Elsevier Science.

solutions to extract pentosans from the prepared cell wall material, as described below.

### 2.2.1.1. Isolation of Cell Wall Material from Wheat Flour

Two methods have been reported for isolating cell wall materials from wheat flour: the flour-water-suspension method and the dough washing method.

*Flour-Water-Suspension Method:* Flour is suspended in water, followed by centrifugation to remove water-soluble material. The insoluble residue is digested with enzyme ($\alpha$-amylase, EC 3.22.1.1, or thermostable amyloglucosidase, EC 3.2.1.2), followed by dialysis or centrifugation to obtain purified cell wall material, as detailed in Figure 4.4 [21–28].

*Dough Washing Procedure:* Dough is prepared by mixing 5 kg of flour and 1.5 L of distilled water in a dough mixer. After 30 and 50 sec, respectively, 760 and 380 mL of distilled water are added. The dough is kneaded for 70 min, then additional distilled water is added at a flow rate of 100 mL/min for 1.5 min. The dough is kneaded for one more minute to allow the water to be absorbed. After this addition and absorption process is repeated five times, the dough is diluted with 10 L distilled water [29]. The slurry is pumped onto a vibrating sieve with a stack of five sieves (250, 125, 90, 50 and 32 $\mu$m, respectively). The fractions on the 32 and 50 $\mu$m sieves are combined as crude cell wall slurry. The residual starch is removed by incubating the slurry with $\alpha$-amylase at pH 6.5, 30°C for 30 hr. After centrifugation, the resulting residue is extracted with hot water (2 L, 63°C) four times and freeze dried to give the water-soluble fraction [30]; whereas, the water-insoluble cell wall material is recovered by freeze drying. It is important to check that the $\alpha$-amylase or any other enzymes used do not degrade the arabinoxylans.

### 2.2.1.2. Extraction of Water-Insoluble Pentosans (WIP) from Cell Wall Material

Once the cell wall material is prepared, the water-insoluble pentosans (WIP) can be extracted under alkaline conditions. Arabinoxylans are usually

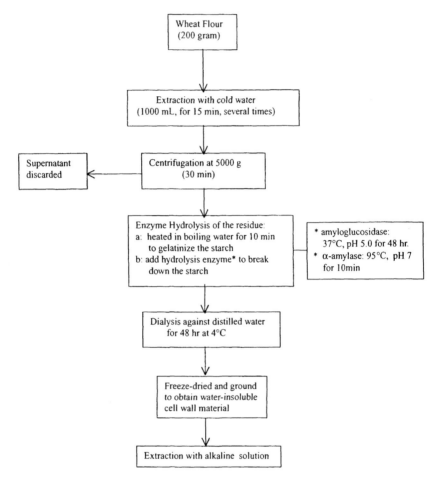

**Figure 4.4** Flowchart of preparation of water-insoluble pentosans from wheat flour.

co-extracted with $(1\rightarrow3)(1\rightarrow4)$-β-D-glucan by NaOH solutions. The solubility of WIP increased with increasing concentrations of alkali, as shown in Figure 4.5 [22]. A sequential extraction of increased solvent strength is generally used to maximize the yield. Solvents used in sequential extractions include chelating agent (e.g., EDTA), diluted $Na_2CO_3$, 1 M KOH or NaOH, 4 M KOH and 4 M KOH with boric acid. Stronger alkali gives a higher yield of total non-starch polysaccharides, however, it frequently causes depolymerization [31].

A saturated solution of $Ba(OH)_2$ containing $NaBH_4$ can selectively extract arabinoxylans from cell wall materials [32]. Composition and molecular weight

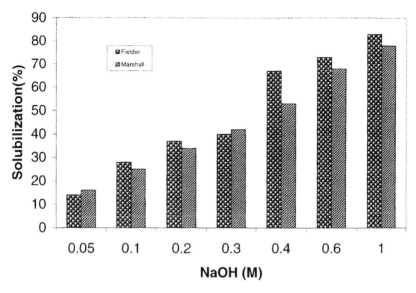

**Figure 4.5** Effect of sodium hydroxide concentration on solubility of water-insoluble pentosans (WIP) from two wheat cultivars. Reprinted from Reference [22].

of arabinoxylan extract using saturated $Ba(OH)_2$ solution is shown in Table 4.6. Although the selective extraction of pentosans with saturated barium hydroxide solutions containing 0.26 M sodium borohydride has already been used on different cell wall materials, the mechanism of its selectivity is still not clear. Generally, a higher concentration of hydroxyl ions will favor a higher yield. Hydroxyl ions are believed to cause swelling of cellulose, disruption of hydrogen bonds between cellulose and hemicelluloses and hydrolysis of esters connected to hemicellulose. The latter may release the arabinoxylans from cross-links of esterified phenolic acids and lignins [33]. Bivalent cations, such as $Ba^{2+}$, are considered to be responsible for the unextractability of β-D-glucans. The solubility of barium hydroxide of 0.23 M at $2°C$ gives a $Ba^{2+}$ concentration of 0.1 M that appears to be sufficient to keep the β-D-glucans insoluble. The presence of 0.26 M sodium borohydride was considered to be jointly responsible for high selectivity because the same concentration of barium hydroxide but significantly lower concentrations of sodium borohydride (0.02 M) solutions extracted more β-D-glucan [16]. In these studies, pretreatment of the water-unextractable cell wall materials to remove or loosen the structures did not improve the yield. Increased temperature can significantly increase the yield of pentosans, but with decreased selectivity; one should also bear in mind that elevated temperature will cause the depolymerization of the arabinoxylans [31].

TABLE 4.6. Yield, Monosaccharide Composition and Molecular Weight of Water Unextractable Cell Wall Materials (WUS) and Pretreated WUS from Wheat Bran after Autoclave, Peroxide and Chlorite Treatment.

| | Yield[a] | Total Sugar Content[b] | Molar Composition[c] | | | | | | | | |
|---|---|---|---|---|---|---|---|---|---|---|---|
| | | | Ara | Xyl | Man | Gal | Glc | AUA | Ara/Xyl | $Mw$[d] | $Mw/Mn$[e] |
| Untreated | | | | | | | | | | | |
| WUS | 100 | 78.9 | 27.7 | 40.7 | 0.2 | 1.1 | 27.0 | 3.3 | 0.68 | $3.9 \times 10^5$ | |
| BE | 19.4 | 87.0 | 41.1 | 53.2 | 0 | 0.9 | 0.8 | 3.9 | 0.77 | | 1.7 |
| Autoclave treated | | | | | | | | | | | |
| AR | 87.5 (100) | 82.7 | 27.3 | 44.6 | 0 | 1.2 | 22.9 | 3.9 | 0.61 | $3.8 \times 10^5$ | |
| BE | 21.6 (24.7) | 86.8 | 42.0 | 53.2 | 0 | 0.9 | 0.4 | 3.5 | 0.79 | | 1.8 |
| Peroxide treated | | | | | | | | | | | |
| PR | 65.4 (100) | 82.5 | 23.8 | 40.3 | 0.2 | 1.0 | 31.1 | 3.6 | 0.59 | $4.3 \times 10^5$ | |
| BE | 16.5 (25.2) | 92.4 | 45.8 | 48.3 | 0 | 1.3 | 0.9 | 3.7 | 0.95 | | 1.8 |
| Chlorite treated | | | | | | | | | | | |
| CR | 75.6 (100) | 78.4 | 29.3 | 43.1 | 0.1 | 0.8 | 24.3 | 2.5 | 0.68 | $5.7 \times 10^5$ | |
| BE | 23.5 (31.1) | 84.2 | 41.8 | 50.0 | 0 | 1.4 | 1.9 | 5.0 | 0.84 | | 2.6 |

[a] Expressed as weight percentage (dm) of WUS. The figures in parentheses represent the yield based on AR, PR or CR.
[b] Neutral sugars + uronic acids expressed as weight percentage (dm) of each fraction.
[c] Expressed as percentage (mole per 100 mole).
[d] $Mw$ = weight average molecular weight.
[e] $Mn$ = number average molecular weight.
AR, PR, CR are residues from autoclave, peroxide, and chlorite treatments, respectively. BE: barium hydroxide extract.
Reprinted from Reference [32] with permission from the *Journal of Cereal Science*. © 1996 Academic Press Ltd.

## 2.2.2. Extraction of Water-Insoluble Pentosans (WIP) from Cereal Bran

The extraction procedure to extract water-insoluble pentosans from wheat bran is similar to that used to extract them from flour. Bran samples are first treated with ethanol/water at a higher temperature to remove lipophilic components and other small molecules and to simultaneously deactivate any endogenous enzymes present. The solvent-treated bran is then subjected to starch gelatinization, followed by digestion with amyloglucosidase or other enzymes to depolymerize the starch. The WIP/arabinoxylan is extracted from the destarched bran using alkaline solutions. A complete fractionation of wheat bran non-starch polysaccharides reported by Brillouet and Mercier is described in Figure 4.6, while the monosaccharide compositions of the fractions are presented in Table 4.7 [34].

A saturated solution of $Ba(OH)_2$ containing $NaBH_4$ is also applicable to selectively extract arabinoxylans from cereal brans [32,35]. It appeared that wheat bran pentosans are more difficult to extract. DuPond and Selvendran [36] reported that only 35% of cell wall material isolated from wheat bran can be solubilized by 4.0 M NaOH. This is because wheat bran contains lignified cell walls in which phenolics and lignin form alkali-stable ester linkages with the polysaccharides. In other words, complete solubilization of polymeric carbohydrates requires delignification of the cell wall material before alkali extraction. Because formation of covalent cross-links between adjacent arabinoxylan chains involves ferulic acid residues, the content of phenolics in cell wall preparations would be inversely related to the degree of solubilization of these materials [36,37].

Schooneveld-Bergmans [38] used three steps to extract arabinoxylans from commercial wheat bran: (a) isolation of water-unextractable cell wall material

**TABLE 4.7. Composition of Polysaccharidic Fractions Obtained by Procedures Described in Figure 4.6.**

| | Non-Starchy Polysaccharide Constituent (%)[a] | | | Uronic Acids Content (%) | Protein Content (%) |
|---|---|---|---|---|---|
| | Xyl | Ara | Glu | | |
| Fraction 1 | 6.6 | 4.6 | 3.2 | 0.9 | 83.0 |
| Fraction 2 | 2.4 | 2.0 | 4.7 | 1.2 | 90.0 |
| Fraction 3 | 14.9 | 8.7 | 11.0 | 3.2 | 60.0 |
| Fraction 4 | 56.6 | 10.1 | 14.2 | 1.4 | 3.8 |
| Fraction 5 | 43.0 | 37.1 | 9.4 | 4.3 | 3.3 |

[a]Percentage of corresponding fraction.
Reprinted from Reference [34] with permission from the *Journal of the Science of Food and Agriculture.*

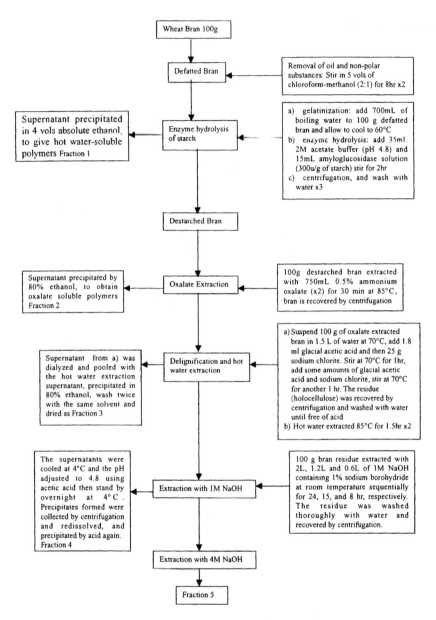

**Figure 4.6** Scheme of the fractionation procedure of wheat bran carbohydrates. Reprinted from Reference [34] with permission from the *Journal of the Science of Food and Agriculture*. © John Wiley & Sons Limited.

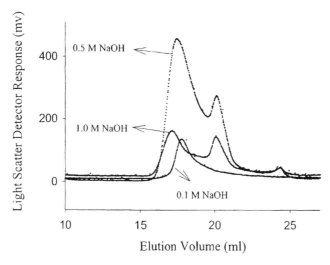

**Figure 4.7** Effect of NaOH concentration on the molecular weight of non-starch polysaccharides (NSP) (mixture of pentosan and β-D-glucan). Reprinted from Reference [31].

(WUS); (b) pretreatment of WUS by autoclave, alkaline peroxide delignification and chlorite delignification; and (c) selective extraction of arabinoxylans with saturated barium hydroxide. The pretreatment of WUS resulted in loss of material, especially lignin-like material, protein and carbohydrates, but did not demonstrate significant improvement on the yield of the extracted arabinoxylans by saturated barium hydroxide, as shown in Table 4.6. An increase in extraction temperature improved the yield from 29% at 20°C to 50% at 95°C, however, significant degradation of the polymer is expected. A study carried out in our laboratory indicated that the extraction yield increases with the alkali strength, but strong alkali (e.g., 1.0 M NaOH) depolymerized wheat bran arabinoxylans even at room temperature (25°C) [31], as demonstrated in Figure 4.7 [31]. However, a 0.5 M NaOH solution did not depolymerize the arabinoxylans significantly, therefore, it was used to extract the non-starch polysaccharides from preprocessed wheat bran with a yield of 18.5% [31].

A sequential procedure was reported by Hromádková et al. to extract WIP from rye bran [39]. Four alkaline concentrations were used to give four fractions of pentosans, as shown in Figure 4.8. Unlike the chlorite delignification method, this method isolated rye bran pentosans without treatment with acid, therefore, possible loss of the arabinose residues was avoided. Native bran is first treated with hot, diluted alkali under nitrogen. Some of the alkali-labile lignin- or phenolic acid-hemicellulose bonds are cleaved during this treatment, thereby solublizing the pentosans. It is believed that alkaline degradation of the arabinoxylan chains combined with losses of arabinosyl side chains is

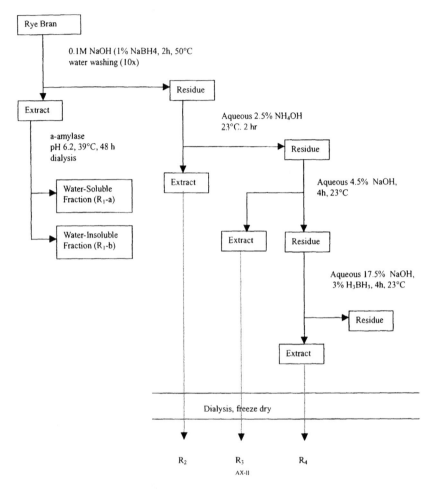

**Figure 4.8** Fractional extraction of rye bran pentosans without chlorite delignification. Adapted from Reference [41] with permission from Elsevier Science.

minimized by adding of sodium borohydride and carrying out the reaction under nitrogen [40].

Most of the highly branched arabinoxylan was extracted during treatment with diluted alkaline and ammonium hydroxide ($R_1$-a, $R_1$-b and R2, respectively). Treatment with 4.5% sodium hydroxide afforded a low branched arabinoxylan ($R_3$ AX-II) that only contained 18 arabinose out of 100 xylose residues [41].

# 3. CHEMISTRY AND STRUCTURAL PROPERTIES OF PENTOSANS

Cereal pentosans are composed of three monosaccharides, i.e., α-L-arabino-furanose (Ara $f$), β-D-xylopyranose (Xyl $p$) and β-D-galactopyranose (Gal $p$). Arabinoxylan, the major component of pentosans, is mostly composed of arabinose and xylose, sometimes with small quantities of ferulic and/or uronic acids, depending on origins and extraction conditions. Much research has been done on wheat and barley arabinoxylans; in contrast, only little attention has been given to arabinogalactans, partially because of their relatively low level and smaller MW. This section focuses on the detailed chemical structure/property of arabinoxylans from different cereals.

## 3.1. STRUCTURE OF WATER-SOLUBLE WHEAT ARABINOXYLAN

The first structure model of water-soluble wheat arabinoxylan was proposed by Perlin and coworkers [42,43]. Recent studies [44–46] not only confirmed the previously proposed structure but also provided solid evidence and details of the structure. Arabinoxylans consist of a backbone of $(1 \rightarrow 4)$ linked β-D-xylopyranose residues to which α-arabinofuranose units are attached through $O$-3 and/or $O$-2,3 positions of the xylose residues [43,47]. The branched xylopyranose residues are not only present as isolated residues in the xylan backbone, but also occur in clusters of two, three or four contiguous residues. The relatively smooth domains of unsubstituted xylose units are also present between the branched regions. Furthermore, the pattern and degree of substitution and distribution of arabinose side chains along the xylan backbone vary with cereal sources and tissue locations in the grain. Proof of the presence of another linkage possibility, i.e., Ara $f$ linked at C-2 of Xyl $p$, was reported by Gruppen et al. [48] and Izydorczyk and Biliaderis [46,49]; this linkage was previously found only in beeswing bran of wheat kernel, rye bran and barley endosperm [35,50,51].

Cereal arabinoxylans are heterogeneous polysaccharides. For example, Hoffmann et al. [52] obtained two arabinoxylan fractions from wheat flour when extracting with cold and warm water. The two fractions are heterogeneous with respect to (a) molecular mass, (b) D-xylopyranose to L-arabinofuranose ratio, (c) average distribution of the arabinose side chains along the xylan backbone and (d) trans-ferulic acid content. The cold-water-soluble arabinoxylans differed from the warm-water-soluble ones in that they have a relatively larger amount of unbranched xylose residues in the backbone, a major group of low molecular mass polymers and the absence of trans-ferulic acid as a non-carbohydrate constituent. The structural heterogeneity of water-soluble wheat arabinoxylan was also demonstrated by Izydorczyk and Biliaderis who

TABLE 4.8. **Ratios of Component Sugars in Arabinoxylans.**

| Variety | Arabinose | Xylose | Glucose |
|---------|-----------|--------|---------|
| HY 355 | 1.00 | 1.89 | 0.32 |
| HY 320 | 1.00 | 1.75 | 0.09 |
| Oslo | 1.00 | 1.52 | 0.04 |
| Glenlea | 1.00 | 1.41 | 0.04 |
| Fielder | 1.00 | 1.58 | 0.14 |
| Norstar | 1.00 | 1.56 | 0.04 |
| Marshall | 1.00 | 1.43 | 0.05 |
| Katepwa | 1.00 | 1.67 | 0.12 |

Reprinted from Izydorczyk et al (1991) with permission [11].

fractionated arabinoxylans into three fractions using a graded ammonium sulphate precipitation technique [53]. The three fractions, F65, F75 and F100, obtained at 65%, 75% and 100% saturation levels of $(NH_4)_2SO_4$, respectively, not only exhibit significant differences in their monosaccharide composition (Table 4.8) [11] but also give different elution profiles from gel filtration chromatography on Sepharose CL-4B (Figure 4.9) [53]. Fraction F65 eluted mostly in the vicinity of the void volume indicating large molecular weight species. In contrast, the elution profile of F100 shifted toward the lower molecular mass end. These structural differences are also clearly demonstrated by the $^{13}C$ NMR spectra, as shown in Figure 4.10. The resonances in the region of 108–110 ppm and 100.8–102.5 ppm are attributed to anomeric carbons of $\alpha$-L-Ara$f$ residues and $\beta$-D-Xyl$p$, respectively. The peaks at 109.4 and 108.7 ppm correspond to C-1s of two $\alpha$-L-Ara$f$ residues linked to a single xylose at C-2 and C-3, respectively. The peak at 108.4 ppm corresponds to C-1 of $\alpha$-L-Ara$f$ linked to a xylose residue at C-3 only. An upfield shoulder of the 109.4 ppm resonance observed in the spectra was suggested to arise from small amounts of $\alpha$-L-Ara$f$ linked to xylose at C-2 only. The resonance at 100.7 ppm is attributed to C-1 of $\beta$-D-Xyl$p$ doubly substituted at C-2 and C-3 with arabinoses. The 101.9 ppm resonance has been assigned to C-1 of unsubstituted xylose residues adjacent to mono- and disubstituted xyloses at their non-reducing end. The signal at 102.4 ppm is assigned to the remaining unsubstituted, as well as the monosubstituted xyloses [52,53]. The relative intensity of signals in the anomeric region of $\beta$-D-Xyl$p$ and $\alpha$-L-Ara$f$ are consistent with those of the C-2, C-3 and C-5 resonances. From these resonances, the relative amounts of di-, mono-, and unsubstituted Xyl$p$ residues and arabinoses linked to single xylose residues at C-3 only, can be deduced.

Substantial intervarietal structural differences were also observed for water-soluble wheat flour pentosans [53–56]. A higher ratio of xylose indicates

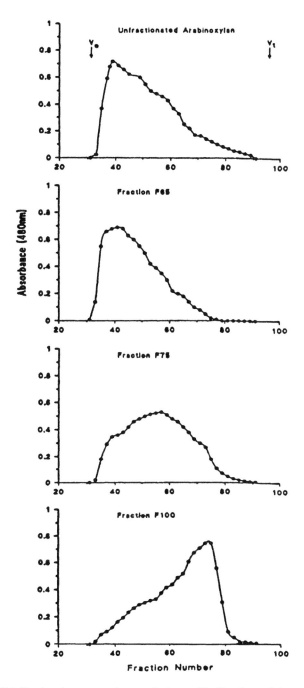

**Figure 4.9** Gel-filtration chromatography on a Sepharose CL-4B column of the unfractionated arabinoxylan (isolated from cv. Katepwa, location Brandon) and fractions obtained by fractional precipitation with $(NH_4)_2SO_4$. $V_t$ and $V_0$ = total and void volumes, respectively. Reprinted from Reference [53] with permission from *Cereal Chemistry*, © 1993.

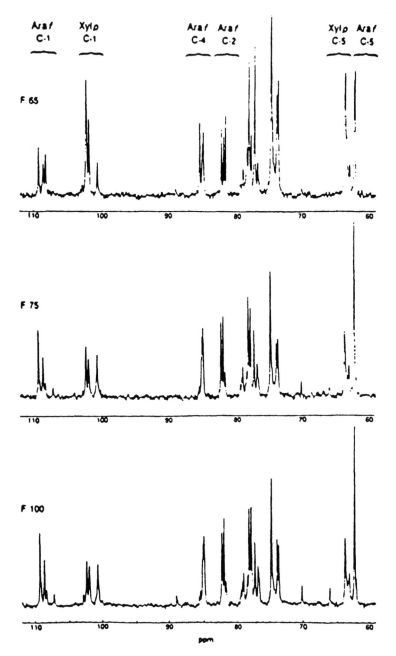

**Figure 4.10** Typical $^{13}$C-nuclear magnetic resonance spectra of arabinoxylan fraction from a presentative arabinoxylan preparation. The chemical shifts were assigned relative to 1,4-dioxane. Reprinted from Reference [53] with permission from *Cereal Chemistry*, © 1993.

a lower possibility of branching. The arabinoxylans of the Canadian western red spring (Katepwa) and prairie spring (HY 355, HY 320) varieties have the lowest degrees of branching; in contrast, the Canada utility variety Glenlea and the hard red spring variety Marshall have the highest degrees of branching [53]. The molecular weight of arabinoxylans from these varieties also varied substantially when accessed by gel filtration chromatography; all arabinoxylans exhibited a broad distribution of molecular weight, as shown in Figure 4.9 [53]. Of the eight wheat varieties examined, higher molecular weights of arabinoxylans were observed from Katepwa, HY 355 and HY 320 compared to those from Norstar, Fielder, Marshall, Oslo and Glenlea.

In order to study the fine structures of water-soluble arabinoxylans, Izydorczyk and Biliaderis [46] further fractionated the arabinoxylans using the graded ammonium sulphate precipitation. Five fractions, F55, F60, F70, F80 and F100, were obtained from ammonium sulphate saturation levels of 55%, 60%, 70%, 80% and 100%, respectively. The relative amount of Xyl$p$ residues doubly substituted at C-2 and C-3 positions with Ara$f$ increased from F55 to F100, whereas the amount of mono- and unsubstituted Xyl$p$ residues decreased in the same order. The linkage compositions of these fractions are shown in Table 4.9 [46].

F55 has three structurally different regions, as shown in Figure 4.11. The first region, $I_{55}$, which accounts for 15% of the polymer, is relatively high in DP (degree of polymerization) and high in terminal arabinose; almost all of these arabinoses are linked to xylose residues doubly substituted at $O$-2 and $O$-3. Fragments corresponding to the first region are of relatively higher molecular weight and are inaccessible to xylanase because of the high frequency of

TABLE 4.9. **Monosaccharide Residue and Linkage Distribution of Arabinoxylan Fractions.**

| | | Molar Composition (%) | | | | |
|---|---|---|---|---|---|---|
| **Alditol Acetate** | **Mode of Linkage** | **F55** | **F60** | **F70** | **F80** | **F100** |
| 2,3,5 Me₃Ara | Ara$f$-(1— | 28.5 | 35.4 | 40.0 | 44.0 | 38.3 |
| 3,5,Me₂Ara | —2)-Ara$f$-(1— | 0.7 | 1.1 | 1.3 | 1.7 | 5.0 |
| 2,5 Me₂Ara | —3)-Ara$f$-(1— | — | 0.4 | 0.6 | 0.6 | 0.7 |
| 2,3 Me₂Ara | —5)-Ara$f$-(1— | 2.2 | 1.4 | 1.8 | 1.8 | 4.4 |
| 2,3,4,Me₃Xyl | Xyl$p$-(1— | 1.1 | 1.2 | 1.8 | 2.4 | 4.8 |
| 2,3,Me₂Xyl | —4)-Xyl$p$-(1— | 36.9 | 37.3 | 30.2 | 25.0 | 20.0 |
| 2 MeXyl + | —3,4)-Xyl$p$-(1— | | | | | |
| 3 MeXyl | —2,4)-Xyl$p$-(1— | 20.8 | 10.4 | 11.4 | 10.5 | 10.3 |
| | | (23.4)[a] | (9.9) | (8.5) | (6) | (4.1) |
| Xyl | —2,3,4)-Xyl$p$-(1— | 9.8 | 12.8 | 12.9 | 14.0 | 16.5 |

[a]Numbers in parentheses are ratios of C-3 to C-2 monosubstituted Xyl$p$.
Reprinted from Reference [46] with permission from Elsevier Science.

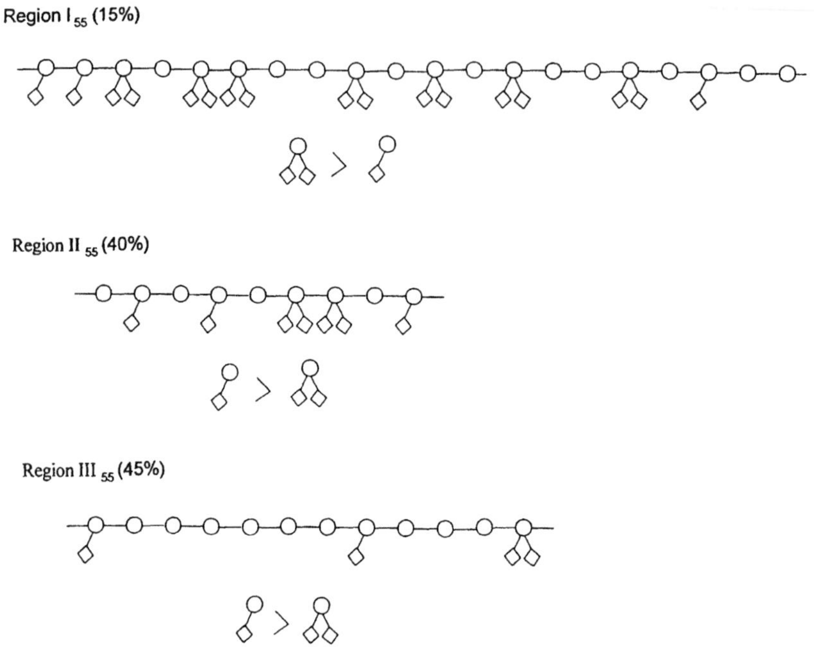

Region I$_{55}$ (15%)

Region II$_{55}$ (40%)

Region III$_{55}$ (45%)

**Figure 4.11** Proposed structure for arabinoxylan fraction F55. Adapted from Reference [46] with permission from Elsevier Science.

arabinose branches. The second region, II$_{55}$, ~40%, contains high amounts of terminal arabinoses linked to xylose residues only at C-3. The third region, III$_{55}$, ~45%, contains higher amounts of contiguously unsubstituted and C-3 monosubstituted xylose residues, and is more accessible to xylanase hydrolysis as evidenced by the presence of small oligosaccharide fragments, such as xylose, xylobiose, xylotriose, arabinosylxylotriose, arabinosylxylotetraose and diarabinosylxylotetraose in the hydrolates [46].

Fraction F100 is built up mainly from highly substituted region I100 (75%), which is structurally similar to the first region of F55. It contains a high amount of terminal arabinoses double linked to O-2 and O-3 of the xylose residues. This fraction also contains a higher amount of O-2 monosubstituted xylose as well as short arabinose side chains. However, there is a lack of Region II$_{100}$ corresponding to Region II$_{55}$. Region III$_{100}$ (about 18%) has a high proportion of unsubstituted xylose residues, however, the amount of disubstituted xylose residues in this region is still higher than that of monosubstituted residues, as shown in Figure 4.12.

A thorough study of the structure of water-soluble wheat flour arabinoxylan was carried out by Hoffmann et al. [13,52]. To gain insight into the distribution

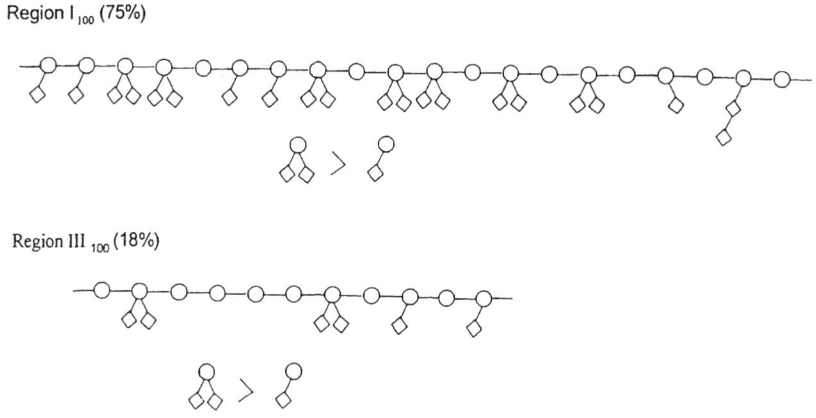

Region I $_{100}$ (75%)

Region III $_{100}$ (18%)

**Figure 4.12** Proposed model for arabinoxylan fraction F100. Adapted from Reference [46] with permission from Elsevier Science.

of arabinose along the xylan backbone, arabinoxylans were hydrolyzed by an endo-1,4-β-D-xylanase to generate oligosaccharides. The structures of these oligosaccharides were elucidated by a series of NMR spectroscopic spectra, including [1]H, [13]C, two-dimensional HOHAHA and two-dimensional ROESY, as demonstrated in Figures 4.13 and 4.14 [13]. The obtained oligosaccharides from enzyme hydrolysis have one or two unbranched xylose residues at the non-reducing end and two or three unbranched xylose residues at the reducing end. These results suggest that the endo-1,4-β-D-xylanase needs a minimum of three contiguous unbranched (1→4)-linked β-D-xylopyranoses for binding and cleavage. Structural features obtained by enzyme hydrolysis combined with two-dimensional HOHAHA and [13]C-[1]H COSY (HMQC) spectra are in agreement with the model proposed by Izydorczyk and Biliaderis [46].

A structural model of water-unextractable arabinoxylan was proposed by Gruppen et al. [45] based on analysis of oligosaccharides released by three purified enzymes from Aspergillus awamori CMI 142717: endo-(1→4)-β-D-xylanase I, endo I-endo-(1→4)-β-D-xylanase III, endo III; and (1→4)-β-D-arabinoxylan arabinofuranohydrolase AXH. The arabinoxylan fragments obtained were characterized and quantified using Bio-Gel P-2 chromatography. A typical Bio-Gel P-2 chromatograph of the arabinoxylan digest is shown in Figure 4.15 [45]. In this model, arabinoxylans are envisaged as containing highly branched regions mostly consisting of tetrameric repeating units of an unsubstituted and a double substituted xylose residue, interlinked with less-branched regions that include subregions containing up to seven contiguous unsubstituted xylose residues. The highly branched regions are enriched in *O*-2,3 disubstituted and *O*-2 monosubstituted, xylosyl residues. The latter is absent in the less substituted regions, as demonstrated in Figure 4.16.

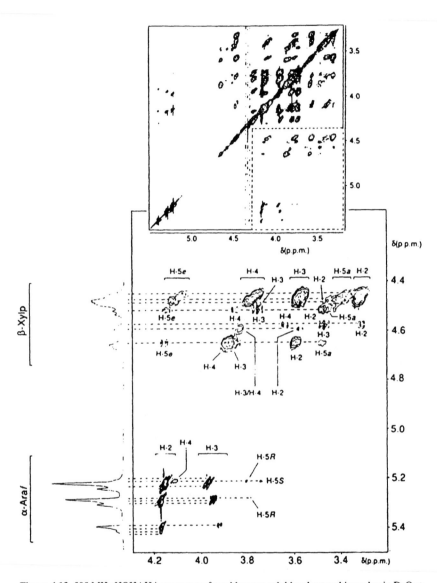

**Figure 4.13** 500 MHz HOHAHA spectrum of a cold-water-soluble wheat arabinoxylan in $D_2O$ at 67°C. A selected region, showing the skeleton-proton resonances along the H-1 tracks, is enlarged and lines are drawn to show scalar coupled networks. The one-dimensional $^1$H-NMR spectrum of the anomeric region is included along the $\omega_1$-axis. Reprinted from Reference [13].

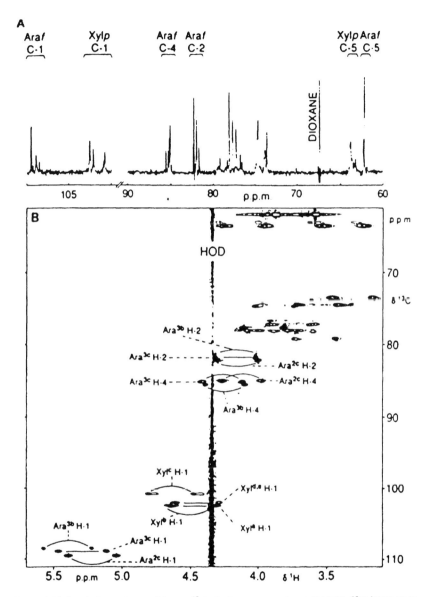

**Figure 4.14** Resolution-enhanced 75 MHz $^{13}$C-NMR spectrum (A) and 500 MHz $^{13}$C-$^1$H HMQC spectrum (B) of a cold-water-soluble wheat arabinoxylan in $D_2O$ at 70° and 67°, respectively. Reprinted from Reference [13].

191

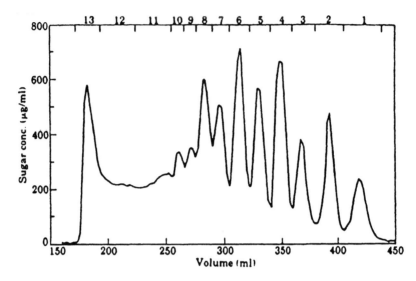

**Figure 4.15** Typical Bio-Gel P-2 chromatogram of arabinoxylan digest with endo(1→4)-β-D-xylanase. Reprinted from Reference [45] with permission from the *Journal of Cereal Science.* © 1993 Academic Press Ltd.

## 3.2. CHEMISTRY AND STRUCTURE OF WHEAT BRAN ARABINOXYLAN

The major constituents of wheat bran are non-starch polysaccharides (41–60%), including arabinoxylan, cellulose and (1→3)(1→4)-β-D-glucan. Of the non-starch polysaccharides, ∼70% are arabinoxylans [57]. In addition to the presence of ferulic acid, there is about 6–9% uronic acids found in the arabinoxylan, including glucuronic acid (70%) and 4-O-methyl glucuronic acid (30%) [36,50]. Therefore, wheat bran arabinoxylans are also mentioned as glucuronoarabinoxylan in the literature [38].

Region A

Region B

**Figure 4.16** Possible structure models proposed for water-unextractable arabinoxylans from wheat flour. Adapted from Reference [45] with permission from the *Journal of Cereal Science.* © 1993 Academic Press Ltd.

Wheat bran arabinoxylans are not soluble in water but can be extracted with alkaline solvents, such as sodium hydroxide and barium hydroxide solutions. A steam explosion method was also used for extracting wheat bran arabinoxylan, however, it results in lower purity and substantial degradation of the polymer. Studies of the structural features of the arabinoxylans are carried out using methylation analyses and degradability by enzymes with a known mode of action [38]. The isolated glucuronoarabinoxylan consists primarily of a lowly substituted population (X/A-ratio of ~5) and a highly substituted population (X/A-ratio ~1). From the methylation analysis data, it was calculated that the lowly substituted glucuronoarabinoxylan comprised ~30% of the total extract, and the branching occurred predominately through monosubstitution at the $O$-3 position of the xylose residues with terminal arabinose. Some disubstitution with arabinose and monosubstitution at the $O$-2 position, probably with glucuronic acid, are also observed. The substituents are randomly distributed and interrupted by stretches of up to six or more contiguously unsubstituted xylose residues as demonstrated by enzymatic degradation studies. The highly substituted glucuronoarabinoxylan comprised about 50% of the total extract [38], and its structural feature is characterized by low X/A-ratio and a relatively large amount of terminal xylose and nonterminal arabinose, as shown in Table 4.10. It is also seen that wheat bran arabinoxylans are very similar to those of rye bran as both consist of lowly and highly substituted xylose residues. The lowly substituted arabinoxylans of wheat bran and rye bran are quite similar to those of wheat flour.

## 3.3. CHEMISTRY AND STRUCTURE OF ARABINOXYLANS FROM OAT, BARLEY AND RYE

*Oat Arabinoxylans:* Water-soluble arabinoxylan only accounts for 3% of oat bran. Hydrolysis of permethylated oat bran arabinoxylan afforded the following sugars that were identified by glc-mass spectroscopy of the derived partially methylated alditol acetates: 2,3,5-tri-$O$-methylarabinose, 2,3-di-$O$- and a mixture of 2-$O$- (mainly) and 3-$O$-methylxylose, together with xylose from doubly substituted residues [58]. A general structure of oat bran arabinoxylan was proposed assuming the same enantiomeric and anomeric configurations as in similar arabinoxylans from other cereals and grasses (Figure 4.17.)

*Barley Arabinoxylans:* In the model for water unextractable barley arabinoxylans, regions of enzyme-resistant polymeric fragments are separated by regions of contiguous unsubstituted xylose residues. The major region consists of isolated unsubstituted residues separated by one or two substituted residues (Figure 4.18) [59]. Blocks of this type are separated by short sequences consisting of two or more unsubstituted xylose residues [Figure 4.18(b)].

TABLE 4.10. Glycosidic Linkage Composition of a Barium Hydroxide Extract from Wheat Bran WUS and some Fractions Derived thereof, Expressed as Percentage (mole per 100 mole) of all Partially Methylated Alditol Acetates Present [38].

| Alditol Acetate of[a] | BE[b] | BE.DU | BE.DB | BE.10 | BE.20 | BE.60 | BE.70 | BE.80s |
|---|---|---|---|---|---|---|---|---|
| 2,3,5-Me$_3$-Ara | 24.6 | 29.4 | 28.9 | 14.1 | 15.8 | 28.5 | 33.4 | 28.6 |
| 3,5-Me$_2$-Ara | 1.6 | 1.1 | 2.3 | 0.2 | 0.2 | 2.5 | 2.6 | 2.0 |
| 2,3-Me$_2$-Ara | 1.3 | 0.9 | 2.4 | 0.3 | 0.5 | 1.0 | 1.0 | 10.6 |
| 2,5-Me$_2$-Ara | 4.7 | 2.1 | 7.9 | 0.7 | 0.7 | 6.8 | 8.0 | 6.5 |
| 5-Me-Ara | 3.4 | 0.9 | 6.6 | 0.6 | 0.2 | 6.9 | 7.3 | 3.9 |
| 2,3,4-Me$_3$-Xyl | 6.6 | 3.1 | 10.7 | 0.9 | 0.8 | 10.3 | 11.4 | 6.8 |
| 2,3-Me$_2$-Xyl | 30.1 | 24.4 | 14.6 | 67.7 | 55.2 | 13.0 | 8.5 | 9.8 |
| 2-Me-Xyl | 10.3 | 8.2 | 10.0 | 11.0 | 10.1 | 9.6 | 9.3 | 10.8 |
| 3-Me-Xyl | 0.9 | 0.7 | 1.7 | 0.4 | 0.3 | 1.6 | 1.8 | 0.5 |
| Xyl | 9.0 | 9.3 | 12.1 | 1.8 | 2.4 | 13.3 | 14.2 | 13.5 |
| 2,3,4,6-Me$_4$-Glc | (1.5)[c] | 0 | 0 | 0 | 0 | 0 | 0 | 0 |
| 2,3,6-Me$_3$-Glc | 4.9 | 16.0 | 1.2 | 1.2 | 10.2 | 4.3 | 1.4 | 3.2 |
| 2,4,6-Me$_3$-Glc | 1.0 | 3.1 | 0.2 | 0 | 2.0 | 0.9 | 0.3 | 0.7 |
| Glc | 0.9 | 0.4 | 0.8 | 0.6 | 0.2 | 0.6 | 0.3 | 2.1 |
| Gal | 0 | 0 | 0 | 0 | 0 | 0.1 | 0.1 | 0.2 |
| Man | 0.7 | 0.4 | 0.6 | 0.5 | 0.4 | 0.6 | 0.5 | 0.8 |

[a]2,3,5-Me$_3$-Ara = 2,3,5-tri-O-methy-1,4-di-O-acetyl-arabinose, etc.
[b]BE: Extracted by saturated barium hydroxide. BE.DU: unbound fraction from DEAE-Sepharose ion exchange chromatography. BE.DB: bound fraction from DEAE-Sepharose ion exchange chromatography. BE.10 to BE.80: fractions obtained by graded ethanol precipitation; 10 to 80 correspond to the concentration of ethanol used in the fractionation.
[c]Determined in carboxyl reduced BE, originally present as glucuronic acid or its 4-O-methyl-derivative.
Reprinted from Reference [38].

**Figure 4.17** Structure model of an oat arabinoxylan. Adapted from Reference [57] with permission from Elsevier Science.

The presence of xylose, xylobiose and xylotriose from xylanase digested products evidenced the presence of the contiguous unsubstituted xylose regions [59]. Despite these differences in cereal arabinoxylans, most of the substituted xylose residues are present in small isolated clusters of singly- and doubly-substituted residues.

*Rye arabinoxylans:* Similar to wheat bran arabinoxylans, rye bran arabinoxylans also contained two types of regions, one containing contiguous double-substituted xylose residues and the other consisting of unsubstituted and *O*-3 monosubstituted xylose residues [41,51]. Highly substituted rye bran arabinoxylan (~5.8%) is water soluble, and the Ara/Xyl ratio is ~0.78. The (1→4)-linked β-D-xylose backbone is ~33% single substituted (2- or 3-), ~26% double substituted and ~41% unsubstituted. Mostly, the α-L-arabinose is attached to the xylan backbone as a single unit, however, small amounts of the arabinose are 2-, 3- or 5-linked.

An arabinoxylan (8.6% of bran) with a low degree of branching was isolated by extraction with 4.5% sodium hydroxide and precipitation of the polysaccharide from the alkaline extract by acidification (AX-I) [4]. This material is water insoluble and contains xylose and arabinose in the ratio of 7:1 with small proportions of D-glucose and uronic acid (Table 4.11). Methylation analysis (Table 4.12) revealed that the backbone of AX-I comprised (1→4)-linked β-D-xylopyranosyl residues with L-arabinofuranosyl groups attached to *O*-3 (*O*-2) of the xylosyl residues. The amounts of terminal arabinose and monosubstituted xylose correspond well, indicating no double substitution in this polymer. However, the molar ratio of the terminal xylose residue to other sugar residues was 1:422, suggesting slight branching of the main chain. The assignment of [13]C-NMR spectrum of AX-I is summarized in Table 4.13 [41].

A water-extractable arabinoxylan-protein complex (AXP) was isolated by extraction with diluted aqueous sodium hydroxide (4%) from delipidated,

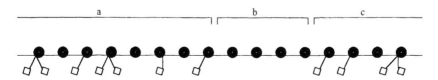

**Figure 4.18** Schematic view of a model for the distribution of substituents over a barley arabinoxylan chain. Reprinted from Reference [58] with permission from Elsevier Science.

TABLE 4.11. **Analytical Data for Arabinoxylans (AX-I and AX-II) Extracted from Rye Bran.**

| | Yield (% of bran) | Rel. Mol %[a] | | | | | $[\alpha]_D^{20}$ (degrees) | Xyl/Ara | $[\eta]$ (dL/g) |
| | | Ara | Xyl | Glc | Gal | Au[b] | | | |
|---|---|---|---|---|---|---|---|---|---|
| AX-I[14] | 8.6 | 12.0 | 85.6 | 1.7 | 0 | 0.7 | −107 | 7.1 | 2.65 |
| AX-II | 8.1 | 13.9 | 81.2 | 3.9 | trace | 1.0 | −105 | 5.8 | 3.20 |

[a]Determined as alditol trifluoroacetates by g.l.c. on OV-225.
[b]4-O-Methyl-D-glucuronic acid.
Reprinted from Reference [41] with permission from Elsevier Science.

destarched, depectinated and delignified rye bran [60]. It contains 1.1% protein and consists of arabinose and xylose in the molar ratio 97:100 with minor amounts of glucose and galactose. Uronic acid, which represents 3.1% of the polysaccharide, was present mainly as D-glucuronic acid and its 4-O-methyl derivative, indicative of acidic xylans known to occur in bran tissues [61]. No phenolic substances detectable as ferulic acid were found, probably because of cleavage of the ester bond during the alkaline extraction step. The molecular weight distribution of AXP showed its heterogeneity: about 70% of AXP were between (MW) $10^4$–$10^6$ and ~26% had MW higher than $10^6$, while the low MW tail ($10^4$) was only about 5%, as shown in Table 4.14 [60].

The protein component of APX was detected by UV absorption at 280 nm in fraction AXP-a only. After incubation with Pronase, the treated samples showed a decrease of average Mw and of the proportion of the high-Mw fraction, depending on the treatment conditions (Table 4.14) [60]. Simultaneously, the UV absorption at 280 nm in the high-MW region was reduced, indicating cleavage of the associated protein. The proteolic treatment of plant gums has been shown

TABLE 4.12. **Linkage Composition of Arabinoxylans from Rye Bran.**

| Methylated Sugar | Mode of Linkage | Molar Ratio |
|---|---|---|
| 2,3,5-Me$_3$-Ara | Ara $f$-(1- | 13.0 |
| 2,3,4-Me$_3$-Xyl | Xyl $p$-(1- | 0.3 |
| 2,3-Me$_2$-Xyl | → 4)-Xyl $p$-(1- | 100 |
| 2- and 3-Me-Xyl | → 3,4)-Xyl $p$-(1- and | 13.5 |
| | → 2,4)-Xyl $p$-(1- | |
| 2,3,6-Me$_3$-Glc | → 4)-Glc $p$-(1- | 2.0 |

Reprinted from Reference [41] with permission from Elsevier Science.

TABLE 4.13. $^{13}$C-NMR Data for Arabinoxylans AX-I and AX-II.

| Ring | C-1 | C-2 | C-3 | C-4 | C-5 |
|------|-----|-----|-----|-----|-----|
| (C) | 101.62 | 72.55 | 73.91 | 75.49 | 53.14 |
| (C′) | 100.95 | 72.25 | 77.31 | 75.78 | 63.55 |
| (C″) | 107.2 | 80.47 | 77.77 | 85.82 | 61.79 |

Reprinted from Reference [41] with permission from Elsevier Science.

to affect their molecular weight distribution, particularly the high-MW fraction, suggesting an association of carbohydrate with protein. However, the absence of xylanase activity needs to be confirmed for the Pronase in order to attribute the breaking down of the high-Mw fraction to the hydrolysis of proteins that linked two or more polysaccharides.

# 4. PHYSICAL AND FUNCTIONAL PROPERTIES OF ARABINOXYLANS

## 4.1. SOLUTION PROPERTIES AND MOLECULAR WEIGHT DISTRIBUTIONS

The solubility of cereal pentosans is largely affected by the arabinosyl substituents on the xylan backbone. In order to examine the effects of substitution on solubility of arabinoxylans, Andrewartha and coworkers prepared a series

TABLE 4.14. Molecular Weight Parameters of AXP and the Pronase-Treated Samples AXP-P1[a] and AXP-P2[b] Estimated by HPGPC on Separon BIO-S 1000.

| Sample | Mw | D[c] | MWD[d] | | |
|--------|-----|------|--------|--------|--------|
| | | | <$10^4$ | $10^4$–$10^6$ | >$10^6$ |
| AXP | 315,000 | 2.33 | 5 | 69 | 26 |
| AXP-P1 | 231,000 | 2.26 | 6 | 79 | 15 |
| AXP-P2 | 235,000 | 2.04 | 6 | 82 | 12 |

[a]Pronase treatment: 37°C, 40 hr.
[b]24°C, 72 hr.
[c]Polydispersity.
[d]Molecular weight distribution (in vol%).
Reprinted from Reference [59] with permission from Elsevier Science.

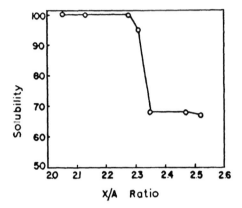

**Figure 4.19** Relationship between the solubility and X/A ratio of the arabinoxylan samples (aX-0 to aX-6) derived from stock arabinoxylan by hydrolysis with α-L-AF (arabinofuranosidase) for increasing time intervals. Reprinted from Reference [62] with permission from Elsevier Science.

of water-soluble arabinoxylans from purified wheat-flour arabinoxylan by partially removing arabinofuranosyl side branches using α-L-arabinofuranosidase [62]. Figure 4.19 shows the progressive change in solubility of the arabinoxylan samples obtained by hydrolysis with α-L-arabinofuranosidase for increasing time intervals as a function of the X/A ratio. As the X/A ratio increased, an abrupt change in solubility occurred between ratios 2.30 to 2.35, and, thereafter, the solubility did not change up to the X/A ratio of 2.55. All of the arabinoxylans remained soluble after incubation with the enzyme for 100 min and showed a slight increase in X/A ratio. However, an insoluble fraction separated after 150 min, and this fraction, as expected, progressively increased in X/A ratio (Figure 4.20) [62]. The soluble fraction, on the other hand, showed a slight decrease in X/A ratio from 100 to 200 min, and thereafter remained constant. The soluble fraction, representing ~65% of the total arabinoxylans, remained soluble throughout the incubation and is apparently not susceptible to the arabinofuranosidase action. The rest of the arabinoxylan is susceptable to enzymatic hydrolysis and, with progressive loss of arabinose residues, becomes insoluble [62]. There are two possibilities that might affect enzyme susceptability. The first possibility is the position of arabinofuranosyl substitution of the xylan. It has been shown that one-third of the arabinose substituents are linked through O-2 and that the remainder are linked through O-3 of the xylose residue. If the arabinofuranosidase specifically hydrolyzed only one type of arabinofuranosyl linkage and this predominated on some chains, then its cleavage would result in specific insolubilization of these chains. The second possibility is the arrangement of the arabinofuranosyl linkages on the xylan backbone. Arabinofuranose residues, present in sequences of contiguous substituted xylosyl residues, may be resistant to hydrolysis. Alternatively, the disubstituted xylosyl

ARABINOXYLANS

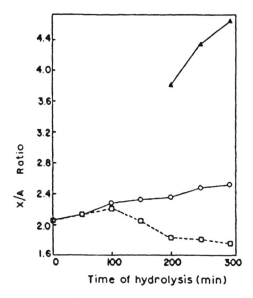

Figure 4.20 Relationship between X/A ratio and time (min) of hydrolysis of the arabinoxylan with α-L-AF (arabinofuranosidase). The total sample (aX-0 to aX-6) derived from the stock arabinoxylan is shown (O), as well as the "soluble fraction" (□) and the "insoluble fraction" (△). Reprinted from Reference [62] with permission from Elsevier Science.

residues may be resistant to hydrolysis, so that molecules with high proportions of one or both of these features would remain unchanged, whereas molecules having fewer of these features would be susceptable and be rendered insoluble. However, these speculations need to be confirmed by structural analysis on these samples using methylation and/or NMR spectroscopy. In addition, the examination of the molecular size of all samples will be very useful in explaining solubility. Although insoluble fractions progressively lose arabinosyl substituents, hydrolysis does not proceed to completion. This may be due, in part, to the increasing insolubility of the arabinoxylan molecules that resulted in lowered accessibility to the enzyme and, in part, to the presence of the resistant features described above.

As the insolubility of $(1 \rightarrow 4)$ $\beta$-D-xylan is due to their ability to aggregate, one can presume that the arabinosyl substituents are possibly responsible for solubilization. The solubilization effect of arabinosyl substituents was not the result of increased hydration but was due to their ability to prevent intermolecular aggregation of unsubstituted xylose residues. Some hydrodynamic characteristics of the soluble portions of arabinoxylan samples are given in Table 4.15 [62]. The degree of hydration can be determined by low-temperature

TABLE 4.15. **Hydrodynamic Characteristics of Arabinoxylan Samples.**

| Sample | X/A Ratio | Intrinsic Viscosity (dL/g) | Hydration (g/g) | Simha Shape-Factor (v) | Axial Ratio (a/b) |
|---|---|---|---|---|---|
| Stock arabinoxylan | 2.00 | 6.12 | 0.23 | 740 | 140 |
| Stock arabinoxylan-boiled | 2.00 | 5.58 | 0.33 | 600 | 125 |
| aX-0 | 2.03 | 5.38 | 0.42 | 530 | 115 |
| aX-1 | 2.11 | 5.14 | 0.51 | 460 | 105 |
| aX-2 | 2.21 | 4.26 | 0.31 | 470 | 105 |
| aX-3 "soluble fraction" | 2.10 | 4.09 | 0.46 | 390 | 96 |
| aX-4 "soluble fraction" | 1.83 | 3.70 | 0.35 | 390 | 98 |
| aX-5 "soluble fraction" | 1.86 | 4.09 | 0.3 | 460 | 105 |
| aX-6 "soluble fraction" | 1.75 | 4.11 | 0.35 | 430 | 100 |

NMR methods. The NMR method measures the amount of water "unfrozen" at $-30°C$ and yields the same values for the degree of hydration of proteins and polysaccharides as obtained by conventional procedures [63,64]. Hydration values reported by Andrewartha et al. [62] apparently were not correlated with the extent of arabinose substitution, possibly due to the large errors that make any small differences in hydration undetectable. It was suggested by the same authors that L-arabinofuranose residues do not contribute to the specific water-binding properties of arabinoxylans, although they could be important for nonspecifically bound water (i.e., osmotic in origin). A linear dependence of intrinsic vicosity with X/A ratio was found within the series of stock arabinoxylan aX-0, aX-1 and aX-2 (Figure 4.21) [62]: the intrinsic viscosity decreased as the arabinose content decreased. Because the arabinoxylans in the series with X/A ratios from 2.0–2.21 have the same MW (aX-2 = 60,000 ± 6000; stock arabinoxylan = 65,000 ± 6500) and the intrinsic viscosity decreased with the series, a decrease in hydration or a change in conformation of arabinoxylan was indicated.

The degree of hydration for these four arabinoxylans is relatively constant, while the axial ratio decreases continuously with a decrease in arabinose content. This indicates that the removal of arabinosyl side chains caused a decrease of asymmetry. The experimentally determined axial ratio of 140 for stock arabinoxylan suggests that this polymer had a very asymmetric solution-conformation, probably assuming an extended rod-like conformation in solution [62].

In order to characterize the MW distribution of an arabinoxylan-protein complex (AXP-P) from rye bran and to establish a Mark-Houwink relationship, gel

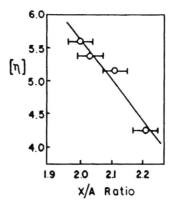

**Figure 4.21** Relationship between the intrinsic viscosity, $[\eta]$ (dL/g), and X/A ratio for stock arabinoxylan-boiled (X/A − 2.00) and α-L AF(arabinofuranosidase)-hydrolyzed samples, aX-0 to aX-2. Reprinted from Reference [62] with permission from Elsevier Science.

permeation chromatography (GPC) coupled with viscosity and light scattering between 30 and 150 degrees was applied by Ebringerova et al. [60]. As shown in Figure 4.22, the polysaccharide was eluted over the entire range of the elution profile indicating a broad size distribution with a small shoulder at the high elution volume end. In the Pronase-treated curve, the high MW species disappeared, and the overall curve shift toward the low MW end. The Mark-Houwink plot of AXP-P showed the same shape with a linear part between MW

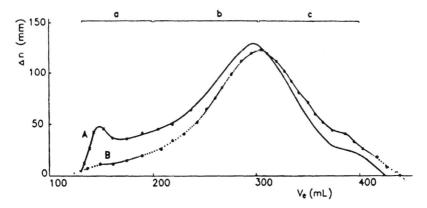

**Figure 4.22** Elution profiles of arabinoxylan-protein complex (AXP, A) and the pronase-treated sample AXP-P(B) on Sepharose CL-2B/Sepharose CL-4B. Reprinted from Reference [59] with permission from Elsevier Science.

TABLE 4.16. Limiting Viscosity and Coil Overlap Parameters
Corresponding to the Diluted and Entangled Domains
of Arabinoxylan Fractions from Wheat Flour.

| | | | | Slope[b] | |
|---|---|---|---|---|---|
| Arabinoxylan Fraction | [η] (dL/g) | c*[a] (g/100 mL) | Coil Overlap (c*[η]) | Diluted Domain | Entangled Domain |
| F60 | 4.70 | 0.26 | 1.24 | 1.13 | 2.19 |
| F70 | 4.20 | 0.31 | 1.30 | 1.12 | 2.04 |
| F80 | 3.16 | 0.38 | 1.20 | 1.07 | 2.00 |
| F95 | 1.90 | nd[c] | nd | nd | nd |

[a]Critical concentration, C*.
[b]Slopes of log $(\eta_{sp})_0$ versus log c[η] on both sides of the critical concentration.
[c]Not determined.
Reprinted from Reference [49] with permission from Elsevier Science.

$10^4$–$10^6$, corresponding to the average MW of the main arabinoxylan of AXP-P with an estimated exponent a = 0.5, characteristic of molecules in unperturbed coil-shaped structures. In contrast, an exponent of a = 0.98 was reported for rye flour pentosans, indicating a lower flexibility of the arabinoxylan chain because of a more extended conformation in solution.

The solution behavior of wheat flour arabinoxylan fractions was reported by Izydorczyk and Biliaderis [49] as summarized in Table 4.16. The zero shear specific viscosity $(\eta_{sp})_0$ was measured at various concentrations of polysaccharides. At low polymer concentrations (dilute domain), $(\eta_{sp})_0$ increased approximately linearly with increasing concentration. But, at higher concentrations (entangled domain), the slopes changed abruptly to a much higher values (Figure 4.23) [49]. This transition is characteristic for polysaccharides whose behavior in solution is influenced by the extent to which individual polymer molecules interact [65]. The abrupt increase in the concentration dependence of $(\eta_{sp})_0$ is attributed to the onset of coil overlap between the polymer chains. The critical concentration (c*), at which the coil overlap occurs, depends on the volume occupied by each molecule. The dimensionless coil overlap parameter (c*[η]), a measure of the total volume occupied by all coils within the polymer solution regardless of their type and molecular weight [49,65], was also obtained from the plot of log$(\eta_{sp})_0$ versus log c[η]. The values of coil overlap parameter for the arabinoxylan was found close to those for relatively stiff and extended polymers like hyaluronate (c*[η] ~2.5) [66], xanthan gum (c*[η] ~0.8) [67] or guar gum (c*[η] ~1.3) [65], which is in accordance with the notion that wheat arabinoxylan assumes a fully extended rod-like shape in solution [62].

*Physicochemical properties of wheat arabinoxylan*

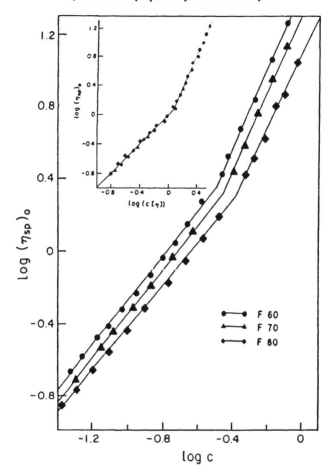

**Figure 4.23** Concentration dependence of zero shear specific viscosity $(\eta_{sp})_0$ for aqueous solutions of arbinoxylan fractions. Inset: zero shear specific viscosity as a function of the overlap parameter $(c[\eta])$ for F60, F70 and F80. Reprinted from Reference [49] with permission from Elsevier Science.

## 4.2. RHEOLOGICAL PROPERTIES OF CEREAL PENTOSANS

### 4.2.1. Viscosity

Substantial differences in limiting viscosities were observed between arabinoxylans (92.81–4.23 dL/g) and arabinogalactoans (0.045–0.062 dL/g) of various wheat flours. Compared to other polysaccharides, the intrinsic viscosities of arabinoxylans are similar to those of guar gum (2.3–6.7 dL/g) and pectin with a low degree of esterification (3.38 dL/g), but they are much more viscous than other neutral polysaccharides such as dextran (0.214 dL/g), beet arabinan (0.195 dL/g) or gum arabic (0.12–0.25 dL/g). The arabinoxylan is the major contributor to the high viscosity of pentosans in aqueous solutions. In contrast, the arabinogalactoan is a component of very low intrinsic viscosity. Previous studies showed lower intrinsic viscosites for wheat flour arabinoxylans (2.5–3.1 dL/g by Marshall et al. [68], and 0.8–3.1 dL/g by D'Appolonia and MacArthur [54]). The differences among these studies might be due to differences in extraction and purification procedures, as well as sample origins, which will essentially result in variations in substitution patterns and molecular weight distributions.

### 4.2.2. Steady Shear Flow

Wheat flour arabinoxylans from a large number of varieties exhibited Newtonian flow behavior up to 1.5% (w/v), except pentosans from cv. Katepwa and HY 355 that exhibited shear thinning flow behavior at high shear rate [68] [Figure 4.24(a)]. At low concentration (e.g., 0.75%), the viscosities of pentosans and arabinoxylans from Katepwa remained relatively constant over a wide range of shear rate (1 to 1000 s$^{-1}$), which is indicative of Newtonian flow behavior (Figure 4.24) [68]. However, at concentrations above 1%, the flow curves exhibited a plateau region at low shear rates as well as a region of decrease of viscosity (shear thinning), typical of a viscoelastic fluid.

The zero shear-rate viscosity of rye and wheat arabinoxylans is almost the same when compared at the same arabinoxylan concentration. In the case of

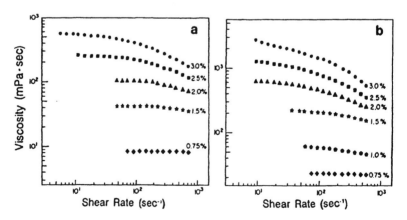

**Figure 4.24** Effect of shear rate on the apparent viscosity of pentosan (a) and arabinoxylan (b) solutions (Katepwa) at concentrations between 0.75 and 3.00% (w/v) at 25 ± 0.1°C. Reprinted from Reference [68].

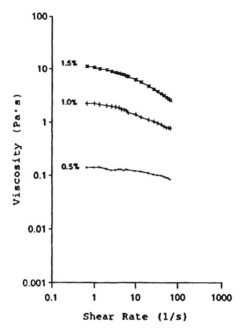

**Figure 4.25** Effect of shear rate on the apparent viscosity of rye arabinoxylans at various polymer concentrations. Reprinted from Reference [69].

wheat arabinoxylans, the onset zero shear-rate at which the apparent viscosity begins to decrease is molecular weight dependent. In the order of decreasing molecular weight, the critical shear rate shifts toward a higher shear rate [19]. Compared with 1.5% wheat arabinoxylan, the viscosity of 1.5% rye arabinoxylan decreases with increasing shear rate (Figure 4.25) [69], which may reflect differences in molecular origin and structure. In fact, a difference of X/A ratio was observed (1.45 for wheat [19] and 2.03 for rye arabinoxylans [69], respectively).

## 4.2.3. Dynamic Oscillatory Measurements

Frequency dependence of storage ($G'$) and loss ($G''$) moduli, also known as the mechanical spectrum, of various wheat arabinoxylans of different molecular sizes were studied by Izydorczyk and Biliaderis [19]. Only fractions with higher molecular weight exhibited elastic properties (Figure 4.26). The frequency at which $G'$ and $G''$ crossed, shifted to higher frequencies at lower concentration and lower molecular weight.

The mechanical spectrum of a non-starch polysaccharide (NSP) from preprocessed wheat bran exhibited viscoelastic properties when freshly prepared

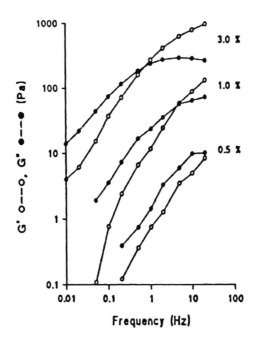

**Figure 4.26** Frequency dependence of storage (G′) and loss (G″) moduli for an arabinoxylan fraction at various concentrations. Reprinted from Reference [19] with permission from the *Journal of Agricultural and Food Chemistry.* © 1992 American Chemical Society.

(Figure 4.27) and exhibited a gel structure when stored at a low temperature (Figure 4.28, 4°C) [31]. At the gel state, the storage modulus G′ predominated the loss modulus G″ over three decades of frequencies (0.01 to 20 Hz). The NSP differed from previously reported wheat pentosans by exhibiting shear thinning flow behavior at low concentrations in water [31] and, more importantly, forming a thermally reversible gel [31]. This unique gelling property will be discussed in detail in the second part of section 4.3.

## 4.3. GELATION OF ARABINOXYLANS

### 4.3.1. Oxidative Gelation

Wheat water-soluble pentosans form viscous solutions in water. But, certain oxidizing agents will cause the formation of a gel. Early observation of oxidative gelation was assumed by a dramatic increase in relative viscosity of a flour suspension upon addition of hydrogen peroxide [70]. Pentosans were found to be the primary responsible factors of the gelling phenomena [71,72]. Oxidative gelation is a unique property of water-soluble wheat-flour pentosans. The

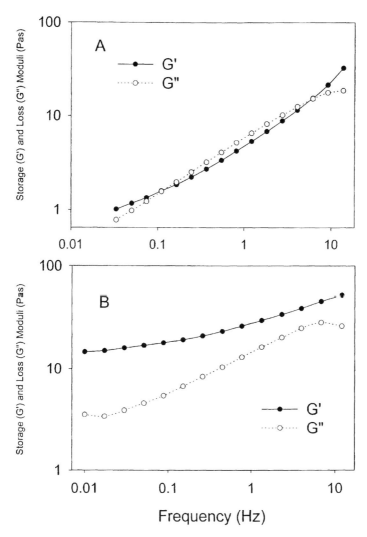

**Figure 4.27** Frequency dependence of storage (G′) and loss (G″) moduli of non-starch polysaccharides (NSP) solution/gel (2.0 %, w/v) at 25°C: A: freshly prepared; B: stored at 4°C for 16 hr. Reprinted from Reference [31].

UV spectrum of pentosans has two peaks, 280 nm and 320 nm, respectively. Kundig et al. showed that the 320 nm peak disappears when oxidizing agents are added [73]. The 320 nm peak is thought to be caused by ferulic acid in the pentosan, thus, ferulic acid is believed to be involved in the oxidative gelation mechanism. Fausch and coworkers evidenced that ferulic acid esterified to the water-soluble pentosans during oxidative gelation [74]. Ferulic acid appears

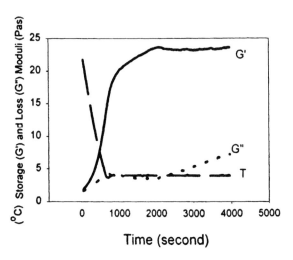

**Figure 4.28** Gel development of 2.0% (w/v) non-starch polysaccharides (NSP) solution dispersion upon cooling as monitored by oscillation measurement at 0.1 Hz. Reprinted from Reference [31].

to have two active centers that could affect cross-linking, presumably because they are involved in the viscosity increase. One is the aromatic nucleus [75], and the other is the activated double bond [76]. To distinguish the two possibilities, Hoseney and Faubion [77] added 250 ppm of fumaric acid, containing the activated double bond but no aromatic nucleus, and of vanillic acid, containing an aromatic nucleus but no activated double bond. Fumaric acid stopped the increase in viscosity, but the vanillic did not. Thus, the author suggested that the active double bond and not the aromatic nucleus is involved in the reaction [77]. However, the same group later corrected themselves and agreed that the aromatic ring of ferulic acid is the center for cross-linking for the pentosans [78], which was originally proposed by Painter and Neukom [79]. Hoseney and Faubion further examined the effect of cysteine based on the observation of Sidhu [76] that $^{14}C$-labeled cysteine was involved in the interaction, and they found that cysteine stopped the viscosity change, suggesting that the sulfhydryl group was involved in the cross-linking. In summary, the oxidation gelation of pentosans involves the aromatic nucleus to form diferulic acid bridges estified to the arabinoxylan fraction of the pentosan (Figure 4.29) [77]. Such covalent grafting of polysaccharide chains would create a multiple cross-linked entity of high molecular weight, which, in the water-soluble system, would be reflected by viscosity increases. In the dough system, the postulated cross-linking would be reflected in changes in dough rheological properties. $H_2O_2$ is effective in causing oxidative gelation. Actually, the active agent is a combination of $H_2O_2$ and the enzyme peroxidase, that is native in the flour.

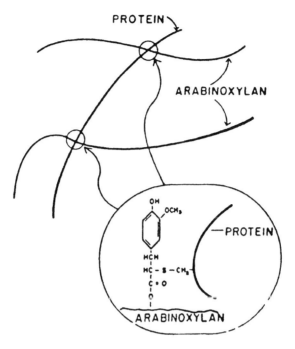

**Figure 4.29** Mechanism of oxidative gelation. Reprinted from Reference [77] with permission from *Cereal Chemistry*, © 1981.

The oxidation gelation can be evaluated by small deformation oscillatory rheological measurement. Figure 4.30 [12] shows the evolution (2% w/v) following the addition of peroxidase (0.11 PU/mL) and $H_2O_2$ (2 ppm) [12]. As described by Izydorczyk et al. [12], the gelation kinetics of pentosan and arabinoxylan solutions assumed a two-phase process: an initial rapid rise in modulus followed by a much slower increase on attainment of a pseudoplateau region. The rapid establishment of a three-dimensional network involves extensive intermolecular cross-linking via feruloyl groups of arabinoxylan molecules. The involvement of feruloyl groups is further evidenced by the disappearance of ferulate residues coinciding with the initial stage of the gelation process (Figure 4.31) [12]. A subsequent slower increase in storage modulus ($G'$) in the pseudoplateau region might reflect additional cross-linking via chain entanglement in the network (Figure 4.30). The effect of oxidative coupling in pentosan solutions was further investigated by gel permeation chromatography on Sepharose CL-4B and Sephacryl S-300. Ferulic acid was associated with the high molecular weight arabinoxylan fraction, which is in accordance with the report by Fincher and Stone [80]. Upon addition of peroxide/$H_2O_2$, a considerable change in the

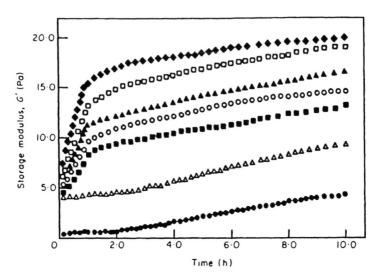

**Figure 4.30** Effect of $H_2O_2$ concentration on time-dependent changes in storage modulus (G′) for 2% (w/v) aqueous solutions of pentosans: the peroxidase activity was 0.11 PU/mL and $H_2O_2$ concentration: 0.3 ppm (●); 1.2 ppm (△); 1.5 ppm (□); 2 ppm (O); 3 ppm (▲); 10 ppm (□); 20 ppm (◆) at 15°C. Data were obtained at 1.0 Hz and 4% strain. Reprinted from Reference [12] with permission from the *Journal of Cereal Science.* © 1990 Academic Press Ltd.

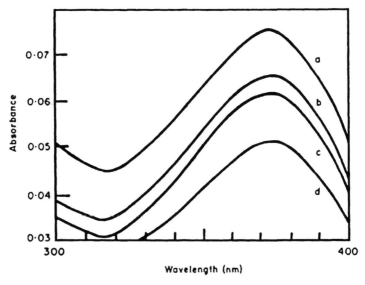

**Figure 4.31** Time-dependent changes in UV spectra of 0.5% (w/v) pentosan solution after addition of peroxidase (0.11 PU/mL) and $H_2O_2$ (0.2 ppm); pentosan solutions (0.5 mL) were mixed with 3.0 mL of 0.07 Glycine-NaOH (pH 10), and the spectra were recorded immediately. (a) Control, 0 time, (b) 45 s, (c) 100 s, (d) 180 s. Reprinted from Reference [12] with permission from the *Journal of Cereal Science.* © 1991 Academic Press Ltd.

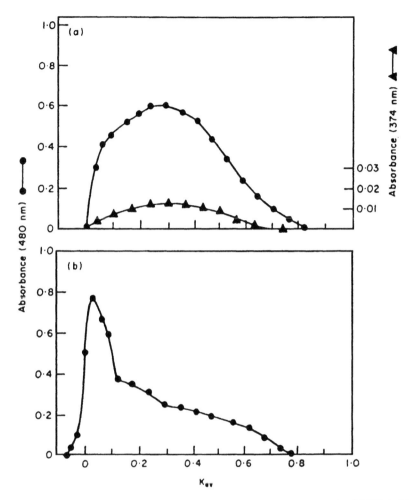

**Figure 4.32** Chromatography on Sepharose CL-4B. (a) native arabinoxylan; (b) 0.375% (w/v) arabinoxylan solution treated with peroxidase (0.11 PU/mL) and $H_2O_2$ (0.45 ppm) at 25°C for 2 hr. Reprinted from Reference [12] with permission from the *Journal of Cereal Science*. © 1990 Academic Press Ltd.

elution profile of the arabinoxylan was observed (Figure 4.32) [12]. Some of the material was eluted in the void volume, while the remainder occurred in the very high molecular weight region. Furthermore, no free feruloyl groups were detected in the eluted fractions. There is no change in the chromatographic profile of arabinogalactan. Although these results are in accordance with rheological data, there is a controversial report based on chromatographic profiles of oxidized pentosan solutions ($Cl_2$ treated) that suggested possible involvement of arabinoxylan-peptide in the cross-linked network [81]. Two of the most important variables in pentosan gelation are polymer and oxidant concentrations, as shown in Figure 4.30 and 4.31, respectively. At low concentrations of $H_2O_2$ (0.3 and 1.2 ppm), G′ increases gradually throughout the entire course of the reaction. At higher concentrations (1.5 and 20 ppm), the cross-linking process is characterized by two stages: a rapid increase in G′ during the first hour, followed by a much lower rate of structure development thereafter (pseudoplateau region). The rate of G′ development and the attained G′ values at the pseudoplateau region increased with increasing oxidant concentration over a 10 hr period. Mechanical spectra of the materials with 0.3 and 20 ppm $H_2O_2$ after a 10 hr reaction period are shown in Figure 4.33 [12]. At 0.3 ppm $H_2O_2$, G′ and G″ were highly dependent on frequency in the range of 0.01 to 10 Hz. The liquid-like character of this system (as manifested by G″) predominated over the elastic component (G′) at all frequencies tested, thus implying a viscous solution. The mechanical spectrum of the sample with 20 ppm $H_2O_2$ is typical of a weak gel, where G′ is greater than G″ at all frequencies and less frequency dependence is observed [12]. The evolution of G′ with time and the extent of

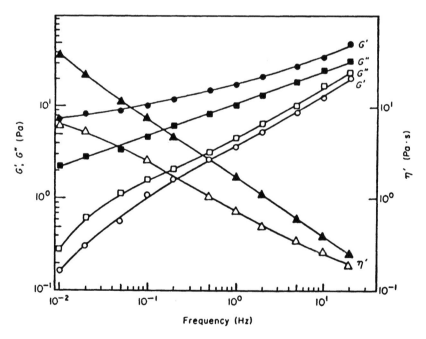

**Figure 4.33** Frequency dependence of storage modulus (G′), loss modulus (G″), and dynamic viscosity (η′) for a 2% (w/v) aqueous solution of pentosans treated with peroxidase (0.11 PU/mL) and $H_2O_2$; 0.3% ppm (open symbols) and 20 ppm (filled symbols) at 15°C. Reprinted from Reference [12] with permission from the *Journal of Cereal Science*. © 1990 Academic Press Ltd.

## GELATION OF WHEAT PENTOSANS

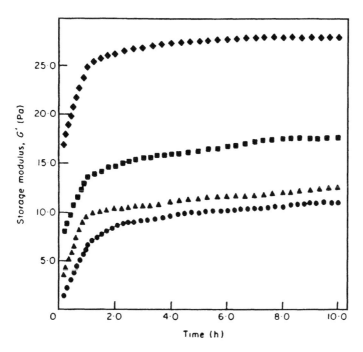

**Figure 4.34** Effect of pentosan concentration on time-dependent changes in storage modulus ($G'$) for pentosan solutions treated with peroxidase 0.11 PU/mL and $H_2O_2$ (1.5 ppm) at 15°C: 1% (●); 1.5% (▲); 2.5% (■); 3.0% (◆). Data were obtained at 1.0 Hz and 4% strain. Reprinted from Reference [12] with permission from the *Journal of Cereal Science.* © 1990 Academic Press Ltd.

gel strength, as manifested by the final values of the modulus (10 hr), increased with increasing polymer concentration (Figure 4.34) [12].

Temperature also has a strong influence on the oxidative gelling mechanism. A stronger gel is formed at 10°C than at 20°C following the usual biphasic pattern. At higher temperatures, the profiles of $G'$ vs. time for pentosan solutions (30 and 40°C) lacked the initial rapid rise in the modulus (Figure 4.35); instead, gradual increases in network rigidity were observed that reached the pseudoplateau values at longer times. The observed $G'$ values of pentosan gels decreased with increasing temperatures in accordance with the known inverse relationship between gel network rigidity and temperature [12]. The temperature may influence the gelation kinetics, presumably, free radical-mediated polymerizations in the sequence of initiation, propagation and termination. The energy of activation of the initiation step is of great importance in determining the overall activation energy of the entire process, and, therefore, temperature

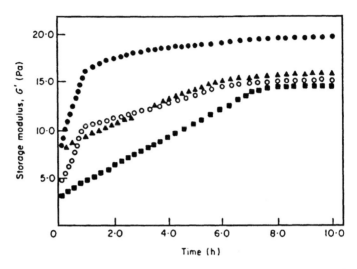

**Figure 4.35** Effect of temperature on time-dependent changes in storage modulus (G′) for 2% (w/v) pentosan solution treated with peroxidase 0.11 PU/mL and $H_2O_2$ (2 ppm); 10°C (●), 20°C (O); 30°C (▲); and 40°C (■). Data were obtained at 1.0 Hz and 4% strain. Reprinted from Reference [12] with permission from the *Journal of Cereal Science*. © 1990 Academic Press Ltd.

variations are expected to influence the reaction rate, i.e., a higher rate with increasing temperature. However, because the initiation step is controlled by the presence of reactive initiator molecules (oxidant), higher temperatures can have adverse effects on the kinetics of the process due to an increased rate of decomposition of the initiator [12].

Oxidative coupling reactions can be induced by different reagents including peroxides, $FeCl_3$, metal oxides ($MnO_2$, $PbO_2$, $HgO$) chromic acid or halohens ($Cl_2$, $Br_2$, $I_2$). However, $KBrO_3$ or $KIO_3$ did not initiate the reaction. The coupling mechanism involves the formation of free phenoxy radicals that are subsequently coupled rapidly and irreversibly to yield mixtures of dimeric and/or polymeric products. The end product of phenolic coupling, therefore, depends on the oxidant used. For example, $(NH_4)_2S_2O_8$ causes pentosan gelation at a much lower rate than $H_2O_2$. This is also corroborated by the relative rates of feruloyl group disappearance for the two. The absorbance at 375 nm, corresponding to free feruloyl groups, decreased very slowly with $(NH_4)_2S_2 4O_8$ (Figure 4.36), compared with $H_2O_2$/peroxidase [12].

There has been some controversy in regard to the effectiveness of $(NH_4)_2S_2O_8$ in promoting gelation of feruloylated wheat pentosans. Hoseney and Faubion [77] observed increases in the viscosity of pentosan solutions; Crowe and Rasper [81] claimed that $(NH_4)_2S_2O_8$ was ineffective as an oxidant of pentosans. In the latter case, however, higher levels of oxidant were used (100 mg/g pentosans vs. 2.5 mg/g pentosans) [12,81].

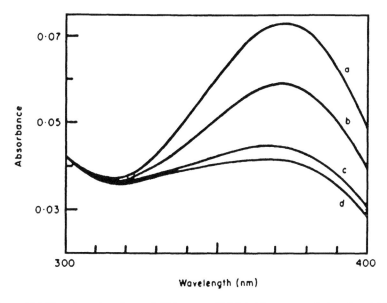

**Figure 4.36**  Time-dependent changes in UV spectra of 0.5% (w/v) pentosan solution after addition of $(NH_4)_2S_2O_8$ (50 ppm); the solutions (0.5 mL) were mixed with 3.0 mL of 0.07 M glycine-sodium hydroxide (pH 10), and the spectra were recorded immediately. (a) Control, 0 time; (b) 1 hr; (c) 3 hr; (d) 5 hr. Reprinted from Reference [12] with permission from the *Journal of Cereal Science.* © 1990 Academic Press Ltd.

It is evident that arabinoxylans extracted from different varieties exhibited a large diversity in gelling capacity. Izydorczyk et al. [12] found that the most rigid gel networks were obtained for arabinoxylans derived from Katepwa (Canada western red spring), followed by those of HY 335, HY 320 and Fielder (Canada western soft spring). Pentosans of Katepwa, HY 355, HY 320 and Fielder showed an initial rapid rise in G′ followed by a plateau region. In contrast, the pentosans from Norstar, Marshall, Glenlea and Oslo exhibited a gradual increase in the G′ throughout the entire reaction time. These findings suggest that the gelling ability of pentosans is related to their structure. Detailed mechanisms of the oxidative gelation are still poorly understood, possibly due to variations in the fine structures of the arabinoxylans, particularly the location of feruloyl groups [68].

Oxidative gelation is also highly pH dependent. 0.55 (w/v) pentosan solution did not gel with $H_2O_2$. However, the gelation speed and the final strength of the gel increased rapidly with a decrease of pH. Below pH 7, the reaction is extremely rapid; at pH 4, the reaction is virtually instantaneous [12].

In summary, the oxidative gelation of arabinoxylans is characterized by a long induction period and a low rate of gel network development. Of all the oxidants examined, $H_2O_2$/peroxidase appears to be the best oxidant system. With regard to the possibility for gelation of water-soluble pentosans in the dough environment, it is well known that all biological systems, including yeast, can generate $H_2O_2$ (readily diffusible through biological membranes) from $O_2$ by dismutation reactions. It is likely, therefore, that $H_2O_2$ is produced during fermenting and proofing. Furthermore, because free radicals have been reported to be present in the water-soluble fraction of flour, activation and coupling (cross-linking) of feruloyl residues could be brought about via other pathways [12].

### 4.3.2. Nonoxidative Gelation

A non-starch polysaccharide (NSP) extracted from preprocessed wheat bran contained arabinoxylan (77%) and β-D-glucan (23%) [83] and some uronic acids [38], and exhibited thermally reversible gelling properties upon cooling to 4°C. The mechanical spectrum of the gel is shown in Figure 4.37. In order to study the contribution of each polymer to the gelling property, two highly specific enzymes were used to depolymerize the individual polymers, namely, xylanase (T. *Viride*, EC 3.2.1.8) for arabinoxylans and lichenase (EC 3.2.1.73) for β-D-glucans. The xylanase completely destroyed the gel structure formed by NSP at 2.0% (w/w). In contrast, lichenase did not destroy the gel but instead enhanced gel strength. The gel development curves, i.e., plots of storage G′ and loss (G″) moduli against time of 2.0% (w/w) NSP and two NSP solutions treated with different enzymes, are shown in Figure 4.38 [84]. The increase in gel strength indicates that the β-D-glucan may not be hydrolyzed completely to oligosaccharides; instead, it may be partially hydrolyzed to a smaller molecular size β-D-glucan. Low molecular weight β-D-glucan has been proved to form gels faster than high molecular weight species [85]. However, it is not clear whether the increased gel strength of NSP aqueous solutions was caused by

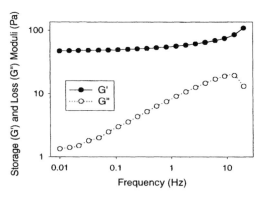

**Figure 4.37** Mechanical spectrum of 2.0% non-starch polysaccharides (NSP) in water at 4°C after cooling at 4°C for 16 hr. Reprinted from Reference [83].

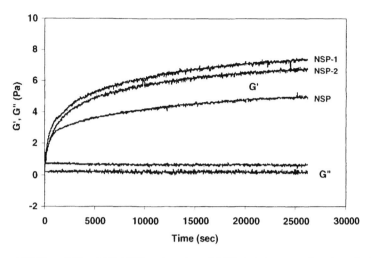

**Figure 4.38** Plot of G′ and G″ (4°C) against time of 2.0% (w/w) non-starch polysaccharides isolated from preprocessed wheat bran. NSP: original; NSP-1, lichenase treated; NSP-2: lichenase and β-glucosidase treated [84].

association between the arabinoxylan and β-D-glucan or by mutual exclusion and the formation of two phases. In order to clarify the role of β-D-glucan in the gel formation of lichenase-treated NSP (NSP-1), β-D-glucosidase, an enzyme used in the standard analytical method to hydrolyze oligosaccharides from lichenase hydrolysis, was added to the solution (NSP-2). This enzyme further hydrolyzes lichenase-released oligosaccharides to produce glucose. After treatment with β-D-glucosidase for 2 hr, the gel strength of NSP-2 was reduced somewhat compared to lichenase-treated NSP-1 but was still higher than the original NSP. This result suggests that the lichenase may not have completely hydrolyzed the β-D-glucan to oligosaccharides that are readily accessible to the β-D-glucosidase. If that is so, the slight decrease of G′ after the addition of β-D-glucosidase might be caused by the removal of glucose from the end of the partially hydrolyzed β-D-glucans that either interacted with arabinoxylans or formed intermolecular junction zones between the β-D-glucans.

The heating curves of NSP and lichenase-treated NSP-1 gels are shown in Figure 4.39. NSP had a melting point at ∼22°C, while the melting point of the lichenase-treated sample (NSP-1) shifted to 25°C with a broader range of transition temperature.

In order to find the contribution of each polysaccharide to the gelling property of NSP, the gelling properties of purified arabinoxylan and β-D-glucan were tested in isolation and in mixtures [84]. As shown in Figure 4.40, 2.0% (w/w) pure arabinoxylan formed a gel under identical conditions to NSP. However, the

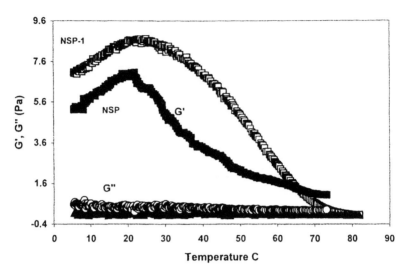

**Figure 4.39** Melting curves of 2.0% (w/w) non-starch polysaccharides isolated from preprocessed wheat bran. NSP: original; NSP-1, lichenase treated [84].

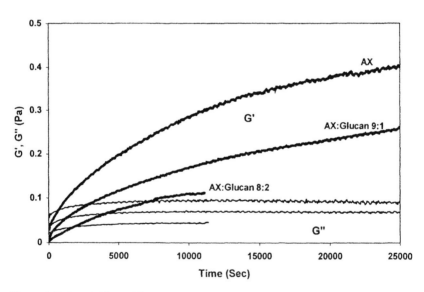

**Figure 4.40** Plot of G′ and G″ (4°C) against time of 2.0% (w/w) arabinoxylans (AX) and their mixtures with β-D-glucans at different ratios [84].

218

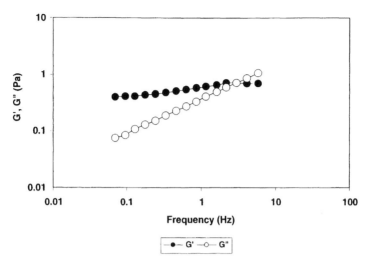

**Figure 4.41** Mechanical spectrum (4°C) of 2.0% (w/w) arabinoxylans isolated from preprocessed wheat bran fraction. Reprinted from Reference [84].

gelation rate and the overall gel strength (G′) are an order of magnitude lower than that of NSP (Figures 4.38 and 4.40). When arabinoxylan and β-D-glucan were mixed in the ratio of 9:1 and 8:2, respectively, gelation rate and gel strength further decreased. It appears that gel strength is related to the amount of AX, and there is no observable synergistic interaction between the arabinoxylan and β-D-glucan after remixing. The mechanical spectrum of AX gel is shown in Figure 4.41. Isolated wheat β-D-glucan did not form a gel within the time period examined, but it formed a strong gel after three days of storage at 4°C, as described in Chapter 3. It can be summarized that arabinoxylan is the component chiefly responsible for the gelling properties of the non-starch polysaccharides extracted from preprocessed wheat bran fraction. However, the role of β-D-glucan cannot be ignored. Although the recomposition of the arabinoxylan and β-D-glucan did not give the gelling properties observed for the NSP, the presence of β-D-glucans significantly increased the gel strength of NSP. The gelation of this NSP involved physical bonding rather than the nonreversible chemical cross-links known for water-soluble wheat pentosans [68,77,86]. This unique nonoxidative gelling property of NSP from preprocessed wheat bran might be related to the presence of uronic acid or other fine structural features [38]. The low degree of substitution of the xylan chain of the arabinoxylan (xylose/arabinose ratio = 3) in this NSP might be responsible for the rheological behavior, although further research is required for a better understanding of the gelling mechanism.

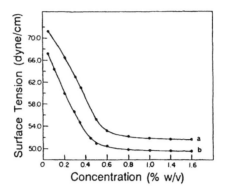

**Figure 4.42** Reduction of the surface tension of water by arabinoxylan (a) and arabinogalactan (b) (HY 355) at various concentrations; the surface tension of distilled water was 72.0 ± 0.5 dyne/cm at 25.0 ± 0.5°C. Reprinted from Reference [68].

## 4.4. SURFACE ACTIVITY AND OTHER FUNCTIONAL PROPERTIES

Water-soluble pentosans exhibited surface activities as demonstrated in Figure 4.42. The surface tension of water decreased rapidly with an increasing concentration of an arabinoxylan and arabinogalactan isolated from wheat flour [68]. All pentosans and arabinoxylans from different varieties were surface active; however, surface activities from different wheat varieties showed considerable variations, as shown in Table 4.17. These pentosans and arabinoxylans also stabilized foams formed by a surface-active protein (bovine serum albumin) (Table 4.18). The addition of arabinoxylans decreased the initial foam volume possibly due to increased viscosity, but it prevented the disruption of gas cells

TABLE 4.17. **Surface Tension of Pentosan, Arabinoxylan and Arabinogalactan Solutions (0.5%, w/v) Isolated from Different Wheat Flour.**

| Variety | Pentosan | Arabinoxylan | Arabinogalactan |
|---------|----------|--------------|-----------------|
| HY 355 | 55.2[a] | 55.4 | 50.8 |
| HY 320 | 58.5 | 56 | 51 |
| Oslo | 48.9 | 48.5 | 50.7 |
| Glenlea | 48.0 | 48.6 | 50.9 |
| Fielder | 47.3 | 44.8 | 50.2 |
| Norstar | 50.8 | 47.3 | 49.6 |
| Marshall | 46.4 | 42 | 46 |
| Katepwa | 50.3 | 49.9 | 50.3 |

[a]Dynes per centimeter; The surface tension of water was 72.0 ± 0.5 dynes/cm at 25.0 ± 0.5°C. Reprinted from Reference [68].

TABLE 4.18. **Stabilization of Foam by Pentosan Constituents Compared to Commercial Gums.**

| | Foam Volume[a] (mL) | | | | |
|---|---|---|---|---|---|
| | | Acidification[b] | | Heating[c] | |
| Polysaccharide | Initial | Immediate | After 10 min | After 3 min | After 5 min |
| Control[d] | $5.0 \pm 0.5$ | $9.5 \pm 0.5$ | $6.5 \pm 0.5$ | $1.0 \pm 0.3$ | . . . [e] |
| **Arabinoxylan[f]** | | | | | |
| HY 355 | $2.3 \pm 0.4$ | $2.4 \pm 0.3$ | $2.4 \pm 0.4$ | $4.5 \pm 0.4$ | $4.0 \pm 0.3$ |
| HY 320 | $2.0 \pm 0.3$ | $2.4 \pm 0.3$ | $2.2 \pm 0.2$ | $4.0 \pm 0.3$ | $4.0 \pm 0.3$ |
| Oslo | $2.3 \pm 0.4$ | $2.5 \pm 0.2$ | $2.4 \pm 0.2$ | $3.0 \pm 0.2$ | $3.0 \pm 0.2$ |
| Glenlea | $1.5 \pm 0.2$ | $2.1 \pm 0.2$ | $1.7 \pm 0.2$ | $2.5 \pm 0.2$ | $2.2 \pm 0.2$ |
| Fielder | $1.5 \pm 0.2$ | $2.6 \pm 0.2$ | $2.3 \pm 0.3$ | $3.0 \pm 0.3$ | $2.8 \pm 0.2$ |
| Norstar | $1.8 \pm 0.2$ | $2.8 \pm 0.3$ | $2.7 \pm 0.2$ | $3.5 \pm 0.2$ | $3.1 \pm 0.2$ |
| Marshall | $1.5 \pm 0.2$ | $2.5 \pm 0.2$ | $2.1 \pm 0.2$ | $3.0 \pm 0.2$ | $2.8 \pm 0.2$ |
| Katepwa | $2.0 \pm 0.2$ | $3.5 \pm 0.2$ | $3.3 \pm 0.2$ | $4.5 \pm 0.3$ | $4.5 \pm 0.3$ |
| **Arabinogalactan[f]** | | | | | |
| HY 355 | $2.8 \pm 0.4$ | $3.7 \pm 0.4$ | $3.0 \pm 0.4$ | $1.5 \pm 0.4$ | . . . |
| HY 320 | $2.2 \pm 0.2$ | $3.5 \pm 0.2$ | $3.4 \pm 0.2$ | $1.0 \pm 0.2$ | $1.0 \pm 0.2$ |
| Oslo | $2.8 \pm 0.4$ | $4.0 \pm 0.4$ | $3.8 \pm 0.2$ | $1.5 \pm 0.4$ | $0.5 \pm 0.2$ |
| Glenlea | $2.5 \pm 0.3$ | $5.0 \pm 0.5$ | $4.5 \pm 0.5$ | $1.0 \pm 0.3$ | . . . |
| Fielder | $2.7 \pm 0.3$ | $3.7 \pm 0.2$ | $3.5 \pm 0.2$ | $1.5 \pm 0.2$ | . . . |
| Norstar | $2.6 \pm 0.3$ | $4.0 \pm 0.3$ | $3.3 \pm 0.3$ | $1.0 \pm 0.2$ | . . . |
| Marshall | $2.8 \pm 0.2$ | $3.8 \pm 0.3$ | $3.5 \pm 0.2$ | $2.0 \pm 0.5$ | . . . |
| Katepwa | $3.5 \pm 0.2$ | $4.5 \pm 0.2$ | $4.0 \pm 0.3$ | $1.5 \pm 0.2$ | $1.0 \pm 0.3$ |
| Xanthan Gum[g] | $2.0 \pm 0.3$ | $4.5 \pm 0.2$ | $3.9 \pm 0.3$ | $8.0 \pm 0.4$ | $7.0 \pm 0.3$ |
| Gum Arabic[g] | $2.2 \pm 0.2$ | $6.5 \pm 0.3$ | $6.0 \pm 0.3$ | $5.0 \pm 0.4$ | $4.5 \pm 0.3$ |

[a]Means $\pm$ SD (n = 3).
[b]Acidification by addition of 0.25 mL of citric acid solution (5%).
[c]Heating in a 95°C water bath.
[d]Control foam formed by mixing well (30 sec) 1 mL of 2% (w/v) bovine serum albumin and 0.25 mL of 5% $NaHCO_3$.
[e]Foam completely disrupted.
[f]Concentration of arabinoxylan and arabinogalactan in foam mixture of 0.2% (w/v).
[g]Concentration of xanthan gum and gum arabic in foam mixture of 0.1% (w/v).
Reprinted from Reference [68].

during thermal expansion of $CO_2$. In contrast, the arabinogalacton failed to stabilize the foam at a higher temperature. The initial formation and expansion of the foam is impeded by the addition of arabinoxylans due to the increase in viscosity of the liquid medium. On the other hand, viscosity and elasticity of the thin film surrounding the gas cells stabilized the foam. Arabinoxylans with high intrinsic viscosity (e.g., Katepwa, HY 355 and HY 320) have an increased ability to stabilize the foams than those of low intrinsic viscosity (Table 4.18). The stabilizing ability demonstrated by wheat arabinoxylans with gluten may play an important role in slowing the diffusion rate of $CO_2$ in the dough during the initial stages of baking. This may affect the loaf volume and final crumb

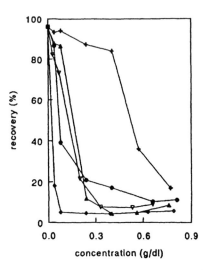

**Figure 4.43** Recovery of paraffin oil from emulsions prepared with the addition of wheat bran glucuronoarabinoxylan or gum arabic to the aqueous phase in varying concentrations. + BE.A; ▲ BE.B; ◆ BE.C; ∇ BE.D; ● gum arabic. Reprinted from Reference [38].

structures: the fineness and homogeneity of the crumb structure is directly re-lated to the collapse and combination of the gas cells during the heating process of baking [68].

Arabinoxylan from wheat bran stabilized emulsion, as shown in Figure 4.43. It is believed that hydrocolloids stabilize emulsions primarily by increasing the viscosity of the continuous phase. BE.A, a low-viscosity arabinoxylan, showed only a low oil recovery at high concentrations. In contrast, BE.C, a high-viscosity arabinoxylan, exhibited a strong ability to stabilize the emulsion even at very low concentrations (0.05 g/dL) [38].

Water absorption during the mixing stage is an important characteristic of wheat flour. The addition of water-insoluble pentosans (WIP) can significantly change the Farinographic water absorption, and the effectiveness of each WIP is concentration dependent, as shown in Table 4.19 [22]. The addition of WIP to Katepwa, HY 320 and Marshall flours markedly increased the water absorp-tion values of the dough. The apparent water-binding capacity of WIP also depends on the chemical and baking qualities of the flour. For example, WIP was added to Katepwa wheat flours with three starch damage levels (28, 36 and 42%). The corresponding water absorption values were 70, 71 and 74%, respectively. These results in conjunction with earlier data (Table 4.19) suggest that supplementation of flour with WIP is a more effective way of increasing the water-binding properties of wheat flour than of increasing the level of damaged starch [22].

TABLE 4.19. **Effects of Added Water-Insoluble Pentosans (WIP) on Farinographic Water Absorption of Wheat Flours.**

| Wheat Flour | WIP Added (%) | Water Absorption (%) |
|---|---|---|
| Katepwa | 0 | 62.3 |
| | 1.0 | 67 |
| | 2.0 | 71.9 |
| | 3.0 | 76.2 |
| | 5.0 | 83.9 |
| Hy 320 | 0 | 55.8 |
| | 1.0 | 61.2 |
| | 2.0 | 65.4 |
| Marshall | 0 | 59.7 |
| | 1.0 | 63.2 |
| | 2.0 | 68.2 |

Reprinted from Reference [22].

## 5. APPLICATIONS AND FUTURE RESEARCH

Pentosans influence baking in at least two ways. First, because of their significant water-holding capacity, they affect the water distribution in the dough; second, water-soluble pentosans exert high viscosity to the dough system, which would help gas retention during fermentation and proof. The ability of wheat dough to retain gas is primarily associated with gluten, but the oxidative gelation of pentosans appeared to affect the loaf volume substantially [77]. In rye dough, gas retention is attributed to the high viscosity of the soluble arabinoxylans [87]. Arabinoxylans in barley and oat are considered anti-nutrients because they slow the nutrient absorption rate in animals. It is desired to isolate these arabinoxylans from barley and oat prior to feeding animals, e.g., using as a gum or dietary fiber. Because pentosans are non-starch polysaccharides, their role in human health is worth exploring. In addition, other non-food applications are also possible.

## 6. REFERENCES

1. Whistler R. L. General comments on future research. In *Progress Biotechnology I: New Approaches to Research on Cereal Carbohydrates*. Hill R., Da M. L., eds. Elsevier, Amsterdam, Oxford, New York, Tokyo. 1985; pp. 413–5.

2. Hashimoto S., Bolte L. C, Pomeranz Y., Shogren M. D. Cereal pentosans—their estimation and significance. 3. Pentosans in abraded grains and milling by-products. *Cereal Chem.* 1987; 64:39–41.

3. Bengtsson S. Structural studies of water-soluble arabinoxylans in rye. Ph.D. Thesis. Swedish University of Agricultural Sciences, Uppsala, Sweden. 1990.

4. Henry R. J. Genetic and environmental variation in the pentosan and beta-glucan contents of barley, and their relation to malting quality. *J. Cereal Sci.* 1986; 4:269–77.

5. Henry R. J. Pentosan and $(1\rightarrow 3),(\rightarrow 4)$-beta-glucan concentrations in endosperm and whole-grain of wheat, barley, oats and rye. *J. Cereal Sci.* 1987; 6:253–8.

6. Hoseney R. C. Functional-properties of pentosans in baked foods. *Food Technology* 1984; 38:114–7.

7. Meuser F., Suckow P. Non-starch polysaccharides. In *Chemistry and Physics of Baking.* Blanshard J. M. V., Frazier P. J., Galliard T., eds. Arrowsmith, Bristol. 1986; pp. 42–61.

8. Bamforth C. W. Barley β-glucans, their role in malting and brewing. *Brew. Dig.* 1982; 57:22–27.

9. Hong B. H., Rubenthaler G. L., Allan R. E. Wheat pentosans. 1. Cultivar variation and relationship to kernel hardness. *Cereal Chem.* 1989; 66:369–73.

10. Hashimoto S., Shogren M. D., Pomeranz Y. Cereal pentosans—their estimation and significance. 1. Pentosans in wheat and milled wheat products. *Cereal Chem.* 1987; 64:30–4.

11. Izydorczyk M., Biliaderis C. G., Bushuk W. Comparison of the structure and composition of water-soluble pentosans from different wheat-varieties. *Cereal Chem.* 1991; 68:139–44.

12. Izydorczyk M. S., Biliaderis C. G., Bushuk W. Oxidative gelation studies of water-soluble pentosans from wheat. *J. Cereal Sci.* 1990; 11:153–69.

13. Hoffmann R. A. Structural characterisation of arabinoxylans from white wheat flour. Ph.D. Thesis. Netherlands, College van Dekanen. 1991.

14. Wood P. J., Weisz J., Blackwell B. A. Molecular characterization of cereal beta-D-glucans—structural-analysis of oat beta-D-glucan and rapid structural evaluation of beta-D-glucans from different sources by high-performance liquid-chromatography of oligosaccharides released by lichenase. *Cereal Chem.* 1991; 68:31–9.

15. Westerlund E., Andersson R., Aman P. Isolation and chemical characterization of water-soluble mixed-linked β-glucans and arainoxylans in oat milling fractions. *Carbohydr. Polym.* 1993; 20:115–23.

16. Wood P. J. Oat β-glucan: structure, location and properties. In *Oats: Chemistry and Technology.* Webster F. H., ed. St. Paul, MN: Am. Assoc. Cereal Chem. 1986. pp. 121–148.

17. Preece I. A., Hobkirk R. Non-starchy polysaccharides of cereal grains III. Higher molecular gums of common cereals. *J. Inst. Brew.* 1953; 59:385–92.

18. Izydorczyk M. S., Macri L. J., MacGregor A. W. Structure and physicochemical properties of barley non-starch polysaccharides—II. Alkali-extractable β-glucans and arabinoxylans. *Carbohydr. Polym.* 1998; 35:259–69.

19. Izydorczyk M. S., Biliaderis C. G. Effect of molecular-size on physical-properties of wheat arabinoxylan. *J. Agric. and Food Chem.* 1992; 40:561–8.

20. Bengtsson S., Aman P. Isolation and chemical characterization of water-soluble arabinoxylans in rye grain. *Carbohydr. Polym.* 1990; 12:267–77.

21. D'Appolonia B. L., MacArthur L. A. Comparison of bran and endosperm pentosans in immature and mature wheat. *Cereal Chem.* 1976; 53:711–8.

22. Michniewicz J., Biliaderis C. G., Bushuk W. Water-insoluble pentosans of wheat: composition and some physical properties. *Cereal Chem.* 1990; 67:434–439.

23. Jelaca S. L., Hlynka I. Water-binding capacity of wheat flour crude pentosans and their relation to mixing characteristics of dough. *Cereal Chem.* 1971; 48:211–222.

24. Cole E. W. Isolation and chromatographic fractionation of hemicelluloses from wheat flour. *Cereal Chem.* 1967; 44:411–416.

25. Yamazaki W. T. The concentration of a factor in soft wheat flours affecting cookie quality. *Cereal Chem.* 1955; 32:26–37.

26. Markwalder H.-U., Neukom H. Diferulic acid as a possible crosslink in hemicelluloses from wheat germ. *Phytochemistry* 1976; 15:836–7.

27. Kim S. K., D'Appolonia B. L. Note on a simplified procedure for the purification of wheat-flour pentosans. *Cereal Chem.* 1976; 53:871.

28. Abdel-Gawad A. S. Isolirung und charakterisierung von pentosanfraktionen aus verschiedenen weizensorten. Ph.D. Thesis. Technische Universitat, Berlin. 1982.

29. Weegels P. L., Marseille J. P., Hamer R. J. Small scale separation of wheat flour in starch and gluten. *Starch/Staerke* 1988; 40:342.

30. Gruppen H., Marseille J. P., Voragen A. G. J, Hamer R. J. On the large-scale isolation of water-insoluble cell wall material from wheat flour. *Cereal Chem.* 1990; 67:512–4.

31. Cui W., Wood P., Weisz J., Beer M. U. Non-starch polysaccharides from pre-processed wheat bran: extraction and physicochemical characterizations. *Cereal Chem.* 1999; 76:129–33.

32. Bergmans M. E. F., Beldman G., Gruppen H., Voragen A.G. J. Optimization of the selective extraction of (glucurono) arabinoxylans from wheat bran—use of barium and calcium hydroxide solution at elevated-temperatures. *J. Cereal Sci.* 1996; 23:235–45.

33. Pomeranz Y. Structure and mineral composition of cereal aleurone cells as shown by scanning electron microscopy. *Cereal Chem.* 1973; 50:504.

34. Brillouet J. M., Mercier C. Fractionation of wheat bran carbohydrates. *J. Sci. Food and Agric.* 1981; 32:243–51.

35. Vietor R. J., Angelino S, A, G. F, Voragen A G. I Structural features of arabinoxylans from barley and malt cell wall material. *J. Cereal Sci.* 1992; 15:213–22.

36. Dupond S. M., Selvendran R. Hemicellulosic polymers from the cell walls of beeswing wheat bran: Part I, Polymers solubilised by alkali at 2°C. *Carbohydr. Res.* 1987; 163:99–113.

37. Neukom H. Chemistry and properties of the non-starchy polysaccharides (NSP) of wheat flour. *Food Sci. and Techn.* 1976; 9:143–148.

38. Schooneveld-Bergmans. M. E. F. Wheat bran glucuronoarabinoxylans (biochemical and physical aspects). Ph.D. Thesis. Dutch Innovation Oriented Programme Carbohydrates (IOP-k), Netherlands. 1997.

39. Hromadkova Z., Ebringerova A. Relation between extractibility and structural heterogeneity of rye bran D-xylan. In *Xalans and Xalanases: Proceedings of an International Symposium, Progress in Biotechnology, Volume 7.* Visser J., Kusters Van Someren M. A., Beldman G., Voragen A. G. J., eds. Elsevier Science Publishers. B.V.; 1992; pp. 395–398.

40. Carpita N. C. Hemicellulosic polymers of cell walls of *Zea coleoptiles. Plant-Physiology.* 1983; 72:515–21.

41. Hromadkova Z., Ebringerova A., Petrakova E., Schraml J. Structural features of a rye-bran arabinoxylan with a low degree of branching. *Carbohydr. Res.* 1987; 163:73–9.

42. Ewald C. W., Perlin A. S. The arrangement of branching in an arabinoxylan from wheat flour. *Can. J. Chem.* 1959; 37:1254–1259.

43. Goldschmid H. R., Perlin, A. S. Interbranch sequences in the wheat arabino-xylan. *Can. J. Chem.* 1963; 41:2272–2277.

44. Hoffmann R. A., Kamerling J. P., Vliegenthart J. F. G. Structural features of a water-soluble arabinoxylan from the endosperm of wheat. *Carbohydr. Res.* 1992; 226(2):303–11.

45. Gruppen H., Kormelink F. J. M., Voragen A. G. J. Water unextractable cell wall material from wheat flour. III. A structural model for arabinoxylans. *J. Cereal Sci.* 1993; 18:111–28.

46. Izydorczyk M. S., Biliaderis C. G. Studies on the structure of wheat-endosperm arabinoxylans. *Carbonhydr. Polym.* 1994; 24(1):61–71.

47. Rotter B. A, Marquaedt R. R, Guenter W., Biliaderis C. G., Newman W. In vitro viscosity measurements of barley extracts as predictors of growth responses in chicks fed

barley-based diets supplemented with fungal enzyme preparation. *Can. J. Anim. Sci.* 1989; 69:431–439.

48. Gruppen H., Hoffmann R. A., Kormelink F. J. M., Voragen A. G. J., Kamerling J. P., Vliegenthart J. F. G. Characterization by H-1-NMR spectroscopy of enzymatically derived oligosaccharides from alkali-extractable wheat-flour arabinoxylan. *Carbohydr. Res.* 1992;233: 45–64.

49. Izydorczyk M. S., Biliaderis C. G. Influence of structure on the physicochemical properties of wheat arabinoxylan. *Carbohydr. Polym.* 1992; 17:237–47.

50. Brillouet J. M., Joseleau J. P. Investigation of the structure of a hetcroxylan from the outer pericarp (beeswing bran) of wheat kernel. *Carbohydr. Res.* 1987; 159:109–26.

51. Ebringerova A., Hromadkova Z., Petrakova E., Hricovini M. Structural features of a water-soluble L-arabino-D-xylan from rye bran. *Carbohydr. Res.* 1990; 198:57–66.

52. Hoffmann R. A., Leeflang B. R., de Barse M. M. J., Kamerling J. P., Vliegenthart J. F. G. Characterization by H-1-NMR spectroscopy of oligosaccharides, derived from arabinoxylans of white endosperm of wheat, that contain the elements -4)(alpha-L-araf-(1-3))-beta-D-xylp-(1→ or -4)(alpha-L-araf-(1-2)) (alpha-L-araf-(1-3))-beta-D-xylp-(1-). *Carbohydr. Res.* 1991; 211:63– 81.

53. Izydorczyk M. S., Biliaderis C. G. Structural heterogeneity of wheat endosperm arabinoxylans. *Cereal Chem.* 1993; 70:641–6.

54. D'Appolonia B. L., MacArthur L. A. Comparison of starch, pentosans, and sugars of some conventional-height and semidwarf hard red spring wheat flours. *Cereal Chem.* 1975; 52: 230.

55. Lineback D. R., Kakuda N., Tsen C. C. Carbohydrate composition of water- soluble pentosans from different types of wheat flours. *J. Food Sci.* 1977; 42:461–467.

56. Ciacco C. F., D'Appolonia B. L. Characterization of pentosans from different wheat flour classes of their gelling capacity. *Cereal Chem.* 1982; 59:96–100.

57. Ralet, M. C., Thibault, J. F., Valle, G. D. Influence of extrusion-cooking on the physicochemical properties of wheat bran. *J. Cereal Sci.* 1990; 11:249–259.

58. Aspinall G. O., Carperter R. C. Structural investigations on the non-starchy polysaccharides of oat bran. *Carbohydr. Polym.* 1984; 4:271–82.

59. Vietor R. J., Kormelink F. J. M., Angelino S. A. G. F., Voragen A. G. J. Substitution patterns of water-unextractable arabinoxylans from barley and malt. *Carbohydr. Polym.* 1994; 24:113–8.

60. Ebringerova A., Hromadkova Z., Berth G. Structural and molecular-properties of a water-soluble arabinoxylan-protein complex isolated from rye bran. *Carbohydr. Res.* 1994; 264:97–109.

61. Wilkie K. C. B. The hemicelluloses of grasses and cereals. *Adv. Carbohydr. Chem. Biochem.* 1979; 36:215–64.

62. Andrewartha K. A., Phillips D. R., Stone B. A. Solution properties of wheat-flour arabinoxylans and enzymatically modified arabinoxylans. *Carbohydr. Res.* 1979; 77:191–204.

63. Cooke R., Kuntz I. D. *Ann. Rev. Biophys. Bioeng.* 1974; 3:95–126.

64. Andrewartha K. A., Brownlee R. T. C., Phillips D. R. *Arch. Biochem. Biophys.* 1978; 185: 423–8.

65. Launay B., Doubblier J. L., Cuvelier G. Flow properties of aqueous solutions and dispersions of polysaccharides. In *Functional Properties of Food Macromolecules.* Mitchell J. R., Ledward D. A., eds. Elsevier Applied Science Publishers, London SP, New York. 1986; pp. 1–78.

66. Morris E. R., Ellis R. Phytate, wheat bran and bioavailability of dietary iron. *Abstracts of Papers of the American Chemical Society* 1981; 181:5.

67. Gravanis G., Milas. M., Rinaudo M., Tinland B. Comparative behavior of the bacterial polysaccharide xanthan and succinoglycan. *Carbohydr. Res.* 1987; 160:259–65.

68. Izydorczyk M., Biliaderis C. G., Bushuk W. Physical properties of water-soluble pentosans from different wheat varieties. *Cereal Chem.* 1991; 68:145–150.

69. Autio K. Functional aspects of cereal cell wall polysaccharides. In *Carbohydrates in Food.* Eliason A.-C., ed. Marcel Dekker Inc., New York, Basel, Hong Kong. 1996; pp. 227–264.

70. Durhum R. K. Effect of hydrogen peroxide on relative viscosity measurements of wheat and flour suspensions. *Cereal Chem.* 1925; 2:297.

71. Baker J. C., Parker H. K., Mize M. D. The pentosans of wheat flour. *Cereal Chem.* 1943; 20:267–280.

72. Baker J. C., Parker H. K., Mize M. D. Supercentrifugates from dough. Cereal Chem. 1946; 23:16.

73. Kundig. W., Neukon H., Deuel H. Untersuchungen uber getreideschleimstoffe. II> uber die gelierung wasseriger von weiaenmehlpentosanen derch oxydationsmittel. *Helv. Chim. Acta.* 1961b; 44:969–976.

74. Fausch H., Kundig W., Neukom H. Ferulic acid as a component of glycoprotein from wheat flour. *Nature.* 1963; 199:287.

75. Neukom H., Markwalder H. U. Oxidative gelation of wheat flour pentosans: a new way of cross-linking polymers. *Cereal Foods World* 1978; 23:374–376.

76. Sidhu J. S., Hoseney R. C., Faubion J. M., Nordin P. Reaction of C14 cysteine with wheat flour water solubles under ultraviolet light. *Cereal Chem.* 1980; 57:380–382.

77. Hoseney R. C., Faubion J. M. A mechanism for the oxidative gelation of wheat- flour water-soluble pentosans. *Cereal Chem.* 1981; 58:421–4.

78. Moore J. N., Cook J. A., Morris D. D., Halushka P. V., Wise W. C. Endotoxin- induced procoagulant activity, eicosanoid synthesis, and tumor-necrosis- factor production by rat peritoneal-macrophages-effect of endotoxin tolerance and glucan. *Circulatory Shock* 1990; 31:281–95.

79. Painter T. J., Neukom H. The mechanism of oxidative gelation of a glycoprotein from wheat flour. Evidence from a model system based upon caffeic acid. *Biochim. Biophys. Acta* 1968; 158:363–381.

80. Fincher G. B., Stone B. A. A water-soluble arainogalactan-peptide from wheat endosperm. *Australian J. Biol. Sci.* 1974; 27:117–132.

81. Crowe N. L., Rasper V. F. The ability of chlorine and chlorine-related oxidants to induce oxidative gelation in wheat-flour pentosans. *J. Cereal Sci.* 1988; 7:283–94.

82. Mazza G., Biliaderis C. G. Functional-properties of flax seed mucilage. *J. Food Sci.* 1989; 54:1302–5.

83. Cui W., Wood P. J., Weisz J., Mullin J. Unique gelling properties of non-starch polysaccharides from pre-processed wheat bran. In Gums and Stabilizers for the Food Industry 9. Williams P. A., Phillips G. O., eds. Royal Chemistry Society, Cambridge, UK. 1998; pp. 34–42.

84. Cui W., Wood P. J., Wang Q. Gelling mechanisms of non-starch polysaccharides (NSP) from pre-processed wheat bran fraction. In Gums and Stabilizers for the Food Industry 10. Williams P. A., Phillips G. O., eds. Royal Chemistry Society, Cambridge, UK. 2000; pp. 156–164.

85. Bohm N., Kulicke W. M. Rheological studies of barley (1-3)(1-4)-β-glucan in concentrated solution: mechanistic and kinetic investigation of the gel formation. *Carbohydr. Res.* 1999; 315:302–11.

86. Geissmann T., Neukom H. On the composition of the water soluble wheat flour pentosans and their oxidative gelation. Lebensm.-Wiss & Techno. 1973; 6:59–62.

87. He H., Hoseney R. C. Gas retention of different cereal flours. *Cereal Chem.* 1991; 68:334–339.

# CHAPTER 5

# Potential Gums from Other Agricultural Resources: Psyllium, Fenugreek, Soybean and Corn Fiber Gums

## 1. INTRODUCTION

IN addition to the polysaccharide gums described in the previous chapters, there are many other agriculture resources that contain significant amounts of water-soluble polysaccharides and have a great potential as stabilizers and thickening and/or gelling agents. However, there are only limited amounts of information available in the literature regarding these materials. This chapter reviews four polysaccharide gums that, in the author's opinion, have great potential of becoming commercial gums.

## 2. PSYLLIUM GUM

Psyllium gum comes from the seed of plants of the *Plantago* genus, which comprises about 200 species of herbs or shrubs widely distributed in the temperate regions of the world [1]. The gum is deposited in the seed coat (husk, hull), and the husk has a long history of medicinal use. Its main applications are as a laxative and for alleviation of large bowel disorder [2]. Recent studies found that psyllium gum significantly lowered plasma low-density lipoprotein (LDL)-cholesterol levels in humans and experimental animals [3,4]. Psyllium husk, which contains about 70% soluble fiber, is currently used as a laxative and dietary fiber supplement in the pharmaceutical and food industries, respectively, and it shows potential for being used as a stabilizer or gelling agent.

## 2.1. EXTRACTION OF PSYLLIUM GUM

Several methods of extracting psyllium gum have been reported. Because the gum is located in the seed coat, it is advantageous to separate the seed coat from the rest of the seed before extraction. The seed coat can be cracked by mechanical pressure or by freezing loosen the hull [5]. The gum can be extracted with boiling water until the seed becomes swollen [6]. The dispersed gum is then separated from cellulose and other insoluble portions by centrifugation. Various extraction conditions gave gum products with different properties [7,8]. Mild and strong alkaline solutions have also been used to extract the polysaccharides. For example, Tomoda and coworkers [9] extracted psyllium seed (100 g) with 0.2% sodium carbonate (1000 mL) at room temperature for 1 hr. After centrifugation, the polysaccharide was obtained by precipitating the supernatant in ethanol, then dialyzing and freeze-drying [9]. Kennedy and coworkers treated the psyllium husk with 1.2 M NaOH, and the extract was subjected to further fractionation [10]. Haque and coworkers [11] used 2.5 M of NaOH to extract the gum from milled husk at ambient temperature for 30 min. The insoluble non-carbohydrate fraction was removed by filtration. The filtrate was neutralized with HCl, dialyzed exhaustedly against deionized water and freeze-dried [11]. The above described extraction procedures have been used in laboratory scales, and the data presented did not appear to be complete. Therefore, studies are needed to optimize the extraction process to obtain a psyllium gum of high yield and high purity and quality, which may lead to the commercial production of psyllium gum.

## 2.2. CHEMICAL COMPOSITION AND STRUCTURES OF PSYLLIUM GUM

Methylation analysis and partial hydrolysis by acid showed that psyllium gum is a highly branched acidic arabinoxylan [9,10]. The xylan backbone having both $(1\rightarrow4)$ and $(1\rightarrow3)$ linkages and the majority of the residues in the xylan backbone are variously substituted at $O$-2 and $O$-3 with arabinose, xylose and an aldobiouronic acid identified as (galactopyranosyluronic acid)-rhamnose, as shown in Figure 5.1 [10]. It is also reported that psyllium polysaccharide contains L-arabinose, D-xylose, D-glucuronic acid and D-galacturonic acid in the ratio of 4.0:10.8:3.3:0.7 and a molecular weight of 1.5 million [9]. Tomoda and coworkers found the presence of $O$-acetyl groups on the polysaccharide at position 2 of about one-fourth of the L-arabinofuranosyl residues, about

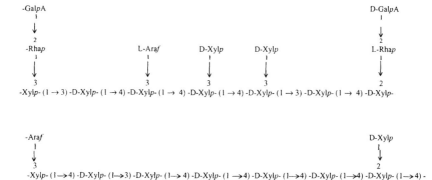

**Figure 5.1** Main structural features of psyllium gum from Plantago ovata Forsk. Reprinted from Reference [10] with permission from Elsevier Science.

two-fifths of the terminal D-xylose residues and about one-fifth of the nonterminal xylose residues [12]. In addition, two new branch structures were reported including an $O$-α-(D-glucuronic acid)-(1→3)-α-L-arabinose and an $O$-α-(D-galacturonic acid)-(1→3)-α-L-arabinose at position 3 as side chains [12], which are different from the structure proposed by Kennedy and coworkers [10].

## 2.3. FUNCTIONAL PROPERTIES OF PSYLLIUM GUM

Psyllium gum does not completely dissolve in water but it swells to a mucilageous dispersion with the general appearance of wallpaper paste [11,13]. The mechanical spectrum and shear-rate dependence of viscosity of 2.0% psyllium gum dispersion are demonstrated in Figure 5.2. It typically exhibits a gel-like structure with elastic response (storage modulus G′) exceeding the viscous response (loss modulus G″) throughout the entire frequency range examined. There is no indication of a constant Newtonian plateau at a lower frequency (ω) for the complex viscosity (η*); instead, log η* decreases with increasing log ω, with a slope of −0.84, which is steeper than the maximum value of −0.76 for disordered polysaccharide cross-linking by topological entanglement [11]. This behavior is similar to that of xanthan gum that generates weak-gel networks by entanglement of rigid, ordered molecular structures [11]. Increasing the concentration of psyllium gum from 1% to 2% gives a significant increase in G′ (from ∼6.5 to 50 Pa at 0.1 Hz) [11]. Steady shear viscosity (η) from rotational measurements on the same sample is significantly lower than η* from

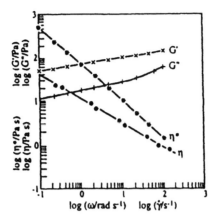

**Figure 5.2** Frequency (ω) dependence of G', G" and η* and shear-rate (γ̇) dependence of viscosity (η) for a 2% dispersion of psyllium gum (isabgol). Reprinted from Reference [13] with permission from Elsevier Science.

small deformation oscillatory measurements (2% strain) at equivalent values of frequency and shear rate (γ̇) (Figure 5.2). This characteristic was defined by Clark and Ross-Murphy [14] as a feature of a weak-gel structure where molecular associations survive low-amplitude oscillation but break down under steady shear. Changes in rheological behavior of psyllium gum in 1% and 2% solutions with temperature are shown in Figure 5.3. At both concentrations, the values recorded at equivalent temperatures on heating and cooling are superimposed,

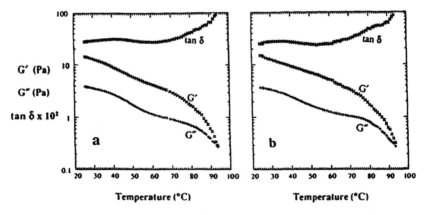

**Figure 5.3** Temperature dependence of G', G" and tan δ for 1% (w/v) psyllium gum (isabgol) on (a) heating and (b) cooling at 1 deg min⁻¹. Measurements were made at 10 rad s⁻¹ and 2% strain. Reprinted from Reference [15] by permission of Academic Press, Inc., © 1993.

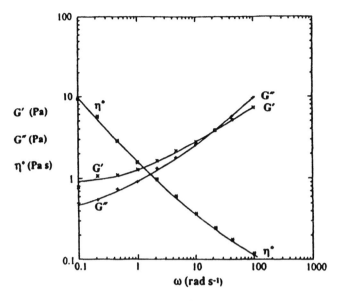

**Figure 5.4** Frequency dependence (2% strain) of G′, G″ and η* for a 2% (w/v) psyllium gum (isabgol) at 91°C. Reprinted from Reference [15] by permission of Academic Press, Inc., © 1993.

indicating that there is no significant thermal hysteresis in the temperature course of rheological change. The most notable feature is a sharp decrease in both moduli above ~80°C, with an associated increase in tan δ (G″/G′). Such behavior corresponds to the onset of gel melting, but the melting process is not completed even after 93°C. This observation is reinforced by high-temperature mechanical spectrum at which the gel-like structure is still evident (Figure 5.4). As shown in Figure 5.5, freshly prepared solutions/dispersions of psyllium gum (1%, w/v) gave flow properties similar to those of disordered coils, with a Newtonian plateau in log η versus log γ̇ at low shear rate. Successive scans between low and high values of shear rates, however, displayed a measured viscosity that demonstrated a progressive shift to higher values at equivalent shear rates. At longer times, solutions of extracted polysaccharide formed cohesive gels that showed obvious syneresis and continued to contract on storage over long periods (to about 30% of their original volume after three months). Substantial (approximately twofold) further contraction was observed on freezing and thawing. However, psyllium gum is stable in high-salt solutions (2.5 M NaCl) formed by neutralization of the alkaline extract over prolonged storage, with no evidence of gelation or precipitation. This was explained as a "salting in" effect, but it is controversial to the normally expected high ionic strength effect that promotes association of charged polysaccharide chains by screening electrostatic repulsion [11].

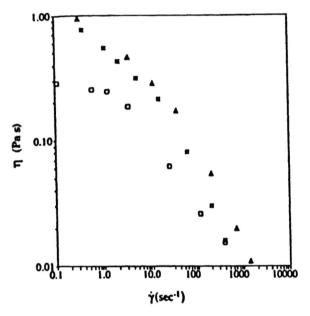

**Figure 5.5** Measured viscosities for a freshly prepared solution of psyllium gum (isabgol) (1% w/v; 25°C ) as the applied shear rate ($\dot{\gamma}$) was increased (□), decreased (■) and again increased (▲) over a period of ~30 min. Reprinted from Reference [15] by permission of Academic Press, Inc., © 1993.

When 1% (w/v) psyllium gum dispersion mixed with 1% (w/v) hydroxy-propylmethylcellulose (HPMC), the weak-gel properties of psyllium gum ($G' > G''$) are clearly evident at low frequencies but are progressively obscured at higher frequencies by the contribution to the overall rheological response from the entanglement-coupling of HPMC, as shown in Figure 5.6 [11,15]. The heating and cooling curves are reversible for the mixture, but the thermal hysteresis from the thermogelation behavior of HPMC is observed (Figure 5.7). A direct comparison of the changes in $G'$ (10 rad s$^{-1}$) on heating for 2% (w/v) psyllium gum, 2% (w/v) HPMC, and a 50/50 mixture of the two (i.e., with both materials present at 1%, w/v) is presented in Figure 5.8. The onset of the steep decline in $G'$ for psyllium gum occurs at about the same temperature (80°C) as the onset of the second, the major wave of increase in $G'$ for HPMC. In the mixed system, the two effects canceled out, with the first, smaller, increase in $G'$ for HPMC similarly balancing the progressive weakening of the psyllium gum network over the same temperature range, to give an essentially constant value of $G'$ at about 55°C. As a comparison, the temperature dependence of $G'$ of gluten is almost identical to that of the psyllium-HPMC mixed system [13]. This similarity established the effectiveness of the two materials when used

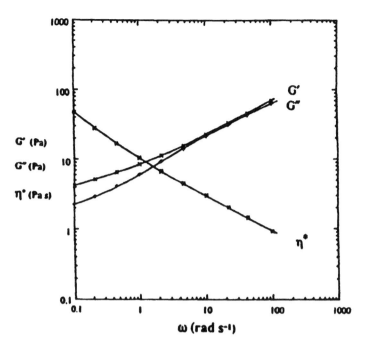

**Figure 5.6** Frequency dependence of G', G'' and $\eta^*$ for 1% (w/v) psyllium gum (isabgol) in combination with 1% (w/v) HPMC, measured at 25°C and 2% strain. Reprinted from Reference [15] by permission of Academic press, Inc., © 1993.

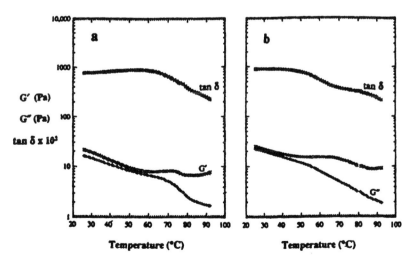

**Figure 5.7** Temperature dependence of G', G'' and tan δ for 1% (w/v) psyllium gum (isabgol) in combination with 1% (w/v) HPMC on (a) heating and (b) cooling at 1 deg min$^{-1}$. Measurements were made at 10 rad s$^{-1}$ and 2% strain. Reprinted from Reference [15] by permission of Academic Press, Inc., © 1993.

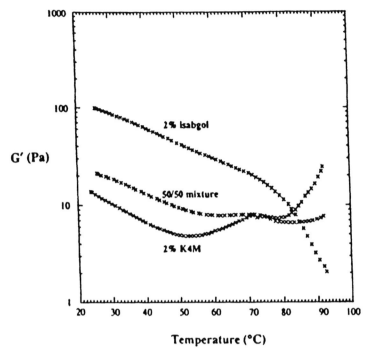

**Figure 5.8** Variation of G' (10 rad s$^{-1}$; 2% strain) on heating (1 deg min$^{-1}$) for 2% (w/v) psyllium gum (isabgol), HPMC (K4M), and a 50/50 mixture of the two samples. Reprinted from Reference [15] by permission of Academic Press, Inc., © 1993.

in combination to replace wheat gluten in breadmaking. As shown in Figure 5.9, the addition of psyllium gum and HPMC (at replacement levels of 2% and 1%, respectively) to rice flour gave a loaf volume close to that of hard wheat control, but neither of them worked alone [13]. The obvious effectiveness of the psyllium-HPMC system appears to arise from the psyllium network stabilizing gas cells as they are formed during proving and preventing them from collapsing during the initial stages of heating in the baking oven, with the cellulosic matrix taking over at a higher temperature [13].

As a dietary fiber, psyllium gum has consistently been shown to lower plasma LDL-cholesterol levels in mildly hypercholesterolemic individuals by 6–20% [16]. Psyllium also effectively reverses dietary-induced hypercholesterolemia in several animal models including the hamster and the African green monkey [17]. A recent study revealed the mechanisms through which the LDL-cholesterol lowering action of psyllium was mediated in the hamster. The LDL-cholesterol lowering action of psyllium in the hamster is carried out through two mechanisms, the major effect is exerted at the level of LDL-cholesterol production. In contrast to their Avicel (microcrystalline cellulose)-fed controls, the hamsters given psyllium had markedly lower plasma total (122.1 vs. 399.4 mg/dL)

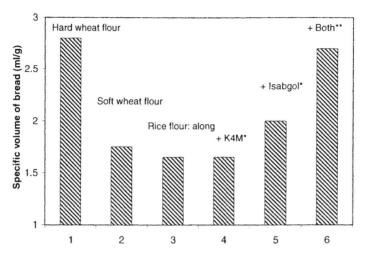

**Figure 5.9** Loaf volumes for bread prepared from wheat flours, and from rice flour, alone or with incorporation of K4M (HPMC) and psyllium gum (isabgol) (*2% replacement level) or both (**2% K4M + 1% isabgol). Reprinted from Reference [13] with permission from Elsevier Science

and LDL-cholesterol (46.0 vs. 143.5 mg/dL) levels. Psyllium feeding also prevented a dramatic increase in hepatic total cholesterol levels (2.6 vs. 16.6 mg/dL) and suppression of hepatic cholesterol synthesis (165.1 vs. 26.1 nmol/h per g) that occurred in the animals given Avicel. Compared to their controls, the psyllium-fed animals also manifested a 44% lower rate of LDL-cholesterol production (167.6 vs. 300.2 ug/h per g bw) and a 2.2-fold higher rate of hepatic LDL clearance (50.1 vs. 22.6 uL/h per g).

## 2.4. APPLICATIONS OF PSYLLIUM GUM

Psyllium gum has been used since ancient times as a demulcent in dysentery, erosion of the intestines, dry coughs, hoarseness, burns, excoriations and inflammations of the eyes [7,8]. The most extensive use of psyllium is as a bulk laxative. When ingested with the proper amount of water, it not only swells and increases the size of the fecal mass, but also, its mucilaginous dispersions have a lubricant action equal to that of oil, without any of oil's disadvantages. Of 24 commercially available bulk laxatives, the swelling and textural properties of psyllium gum are not affected to any extent by 2% sodium chloride, 0.5% hydrochloric acid, 1% sodium hydrogen carbonate, artificial gastric juice and artificial intestinal juice [18]. In artificial intestinal juice, psyllium gum increased in volume 5–14 times compared to locust bean gum (5–10 times) and methylcellulose (16–30 times) in 24 hr [18]. Psyllium gum was also found to be a better laxative than flaxseed gum because it remained in gel form, whereas flaxseed gum did not [19]. The water-holding capacity and gelling property of psyllium gum can also be used to delay and reduce allergic reactions by holding toxins and allergens in the gel structure [19]. Psyllium gum also caused a decrease in skeletal deposition of strontium fed as strontium chloride [19]. With the recent research advances in the area, it is evident that psyllium gum has great potential as a stabilizer and texture improver in the food industry.

## 3. FENUGREEK GUM

Fenugreek (*Trigonella foenum graecum*) is an annual legume traditionally cultivated in Western Asia, Persia, northern India and the Mediterranean region. Its leaves are generally used for preparing curry, and the seeds are used as a condiment or as an anabolism for food and medicine. Fenugreek is a new crop in North America due to a plant breeding program that identified several new cultivars of fenugreek in the prairie provinces of western Canada [20]. Recent research interests in fenugreek are largely due to its high levels of water-soluble fiber (galactomannan or fenugreek gum), which can significantly improve plasma glucose and insulin responses in normal and diabetic subjects [21,22] and exhibit excellent stabilizing properties in oil/water emulsions [23,24].

### 3.1. EXTRACTION OF FENUGREEK GUM

Fenugreek seed contains a very high level of dietary fiber (48.0%). Other components are protein (25.4%), oil (7.9%), saponins (4.8%) and ash (3.9%), as listed in Table 5.1. Of the total dietary fiber, the water-soluble and insoluble fibers are 20% and 28%, respectively. About 6% of starch was also reported in the seed [21,25]. Treatment of either the endosperm or the whole seed with

TABLE 5.1. **Chemical Composition of Fenugreek Seeds.**

| Component | % |
|---|---|
| Moisture | 2.4 |
| Protein | 25.4 |
| Fat | 7.9 |
| Saponins | 4.8 |
| Total dietary fiber | 48.0 |
| Insoluble fiber | 28.0 |
| Soluble fiber | 20.0 |
| Ash | 3.9 |

Reprinted from Reference [25] with permission from Elsevier Science.

water or dilute alkali leads to the extraction of polysaccharide material. The yield of fenugreek gum varied from 14% to 38%, possibly due to differences in variety/source of raw material and extraction methods used [26,27]. In a recent study, Garti and coworkers reported a yield of 20% [28]. In this procedure, fenugreek seeds (125 g) are ground to fine powder (300 mesh) and then extracted with 2 × 100 mL *n*-hexane in a soxhlet until colorless. The *n*-hexane extracted fraction contains mostly lipids, while the solid residue is further extracted with ethanol (200 mL) and then methanol (150 mL). The extract is vacuum evaporated and freeze-dried to produce 7–8 g of saposins. After *n*-hexane and alcohol treatments, the remaining solid residue is dispersed in water (800–1000 mL) to form a viscous aqueous solution. The crude fenugreek gum solution is first centrifuged at 5000 × g to remove non-soluble cellulose, hemicellulose, lignins and part of proteins. The gum solution is further centrifuged at 10,000 × g to precipitate most of the residual proteins and other insoluble fractions [28]. The water extraction process is repeated four times, and all the water-soluble fractions are combined and precipitated with ethanol at a 1:1 weight ratio. The precipitated fenugreek gum (20% of the ground seed) is freeze-dried and ground to a fine powder [28].

## 3.2. CHEMICAL STRUCTURE OF FENUGREEK GUM

Acid hydrolysis and methylation analysis revealed that fenugreek gum is a highly branched galactomannan: the ratio of D-galactose to D-mannose is 5:6 [27]. Similar to other galactomannans, fenugreek galactomannan has a $(1\rightarrow4)$-linked β-D-mannosyl backbone with all the galactose units occupying the terminal position. The molar ratio of terminal galactose, 6-substituted $(1\rightarrow4)$-linked and $(1\rightarrow4)$-linked mannosyl residues is 5:5:1, which suggests that five of every six mannosyl residues are attached by a terminal galactose at the 6 position [27]. The structural feature derived from methylation analysis was also confirmed by periodate oxidation analysis [27]. However, Madar and Shomer [22] suggested that there are two galactomannan fractions in fenugreek seeds: the major fraction was from the endosperm that had a mannose:galactose ratio of 1.5:1, while the second fraction was from seed coat with the mannose:galactose ratio of 1.1:1. In general, fenugreek gum has a similar structural feature to that of other galactomannans. The ratio of mannose to galactose varied with resources, as demonstrated in Figure 5.10.

## 3.3. PHYSICAL PROPERTIES OF FENUGREEK GUM

### 3.3.1. Rheological Properties

There has been little information in the literature on the physical properties of fenugreek gum until recently [23,28,29]. Fenugreek gum solutions exhibit

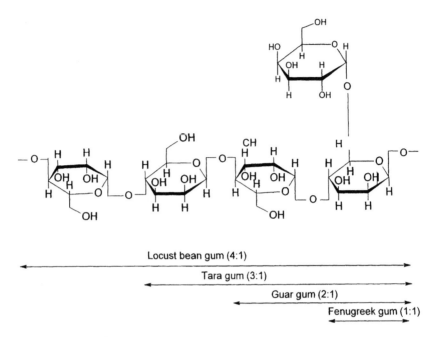

**Figure 5.10** Schematic illustration of the structures of four galactomannans with different galactose/mannose ratios.

similar rheological properties to those of other galactomannans. As shown in Figure 5.11, a Newtonian plateau is observed at a low shear-rate region. When shear rate is increased to a certain extent, the viscosity decreases with the increase of shear rate. However, fenugreek gum has a lower viscosity when compared to guar and locust bean gums at the same concentration and similar molecular weight, especially at low shear rate. The low viscosity of fenugreek gum was explained by its highly substituted structure [28]. A small-strain dynamic oscillatory measurement revealed that fenugreek gum also exhibits viscoelastic fluid behavior comparable to that of locust bean gum, as demonstrated in Figure 5.12, in which the storage modulus G′ was lower than the loss modulus G″ at a low frequency region while the reverse was observed at a high frequency region. Fenugreek gum failed to form a gel when blended with k-carrageenan, while strong gels are formed when guar and locust bean gums are blended with k-carrageenan [28]. In our study, fenugreek gum exhibited synergistic with yellow mustard gum, but to a much lesser extent compared to that of guar and locust bean gums (Chapter 1).

### 3.3.2. Surface and Emulsification Activities

Most polysaccharides are sufficiently hydrophilic, water soluble and rigid, therefore, they are not expected to exhibit significant surface activity [28,30].

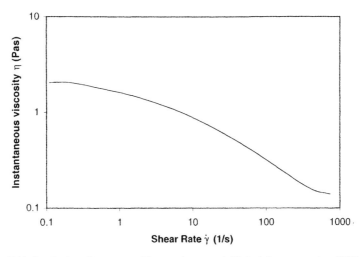

**Figure 5.11** Steady shear flow curve of fenugreek gum at 1.5% (w/w) concentration, 25°C measured on the CVO Rheometer.

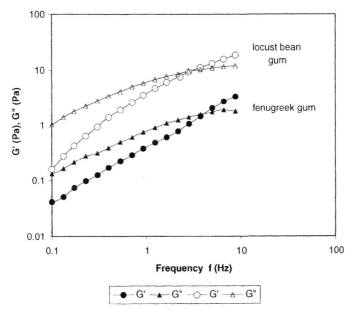

**Figure 5.12** Comparison of mechanical spectra of fenugreek gum and locust bean gum at 1.5% (w/w) polymer concentration, 25°C measured on CVO Rheometer.

Only protein-containing hydrocolloids, e.g., gelatin and gum arabic, are considered to be good steric sabilizers because they contain flexible hydrophobic and hydrophilic groups that behave like emulsifiers. However, commercial galactomannans and native locust bean and guar gums are able to reduce the surface tension of water to ~55 dynes/cm and adsorb/precipitate on oil-water interface and reduce their interfacial tensions [30]. Purified fenugreek gum appeared to be more efficient than guar and locust bean gums in lowering the interfacial free energy. It can reduce the surface tension of water to 42 dynes/cm at 0.7% (w/w) compared to 55 dynes/cm at the same concentration for guar. Fenugreek gum is also more concentration-efficient than gum arabic (60 dynes/cm at 1.0% w/w) or xanthan gum (42 dynes/cm at 1.0% w/w). As shown in Figure 5.13, the curve of decrease in surface tension with increase in concentration of fenugreek gum appeared to be composed of three different stages depending on gum concentration. At a very low concentration (up to ~0.1% w/w), the surface tension reduction was considered to be due to the adsorption of short chain saccharides, which can migrate preferentially to the surface. The second stage (0.1% to 0.5%) probably corresponds to the behavior of macromolecules at the surface. Non-reproducible measurements were obtained for gum concentrations above 0.9% because of the build up of high viscosity in the solution [28]. In order to identify if protein was the active component, fenugreek gum was further purified to eliminate, as much as possible, proteins associated with

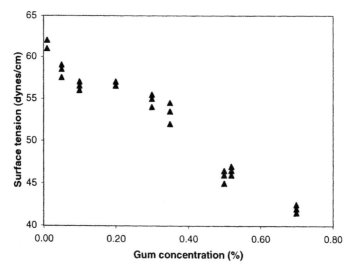

**Figure 5.13** Surface tension of bipurified fenugreek gum aqueous solution as a function of gum concentration. Reprinted from Reference [28] with permission. © 1997 Academic Press Ltd., London, UK.

TABLE 5.2. **Chemical, Physical, Surface, Interfacial and Emulsification Properties of Fenugreek Gum at Various Stages of Purification and Treatment.**

| Fenugreek Gum | Protein (%) | Viscosity (Pas) | Surface Tension $\gamma o$ (Dynes/cm) | Interfacial Tension $\gamma i$ (Dynes/cm) | Average Droplet Size ($\mu$m)[a] |
|---|---|---|---|---|---|
| Purified | 6.0 | 350 | 46 | 26 | 3.0 |
| Bipurified | 2.0 | 250 | 42 | 20 | 2.0 |
| Tripurified | 0.8 | . . . | 40 | 19 | 2.5 |
| Heat treated | 6.0 | . . . | 42 | 26 | 3.0 |

[a]O/W emulsion of 50 g/kg tetradecane, 0.5% (w/w) fenugreek gum.
. . . : not determined.
Reprinted from Reference [28] with permission. © 1997 Academic Press Ltd., London, UK.

the polysaccharides. The further purified fenugreek gum (tripurified) contains only 0.8% protein compared with 6% in the unpurified samples and 2.0% in the bipurified gum [28]. The differences in viscosity, surface tension and interfacial tension of the three grades of purity are demonstrated in Table 5.2. The intrinsic viscosity of the purified solutions was slightly lower than the viscosity of less purified gum, but the surface activity was improved (40 dynes/cm for tripurified gum, 42 dynes/cm for bipurified gum vs. 46 dynes/cm for the purified gum solution). Similar improvement was observed for the $n$-tetradecane/water interfacial tension (20 dynes/cm for the bipurified vs. 19 for the tripurified), as shown in Figure 5.14. It seems that the surface/interfacial activity of fenugreek gum is not affected by the amount of proteins present as impurities although earlier studies demonstrated that other gums lost their activity upon removal of protein from the gums [31–33]. The interfacial tensions of toluene (36 dynes/cm), soybean oil (24 dynes/cm) and castor oil (16 dynes/cm) were reduced by the tripurified fenugreek gum to 7, 3.5 and 2 dynes/cm, respectively. And for any of the other oils, the effect of fenugreek gum on the interfacial tension was greater than that of guar gum [28].

Oil-in water emulsions were prepared with bipurified fenugreek gum at concentrations between 0.1% to 0.9% using a high shear-rate microfluidizer [28]. The droplets were found to be significantly smaller than those obtained from guar gum and are gum concentration dependent (Figure 5.15). At low gum concentrations and partial droplet coverage, the droplets are large and tend to coalesce fast with time. Increase in gum concentration allows better coverage of the droplets resulting in the formation of smaller droplets with only a minor tendency to coalesce. At gum level of 0.9%, the droplets are relatively small (3 $\mu$m) and do not exhibit flocculation or coalesence [28]. The droplet size distribution can be plotted against the gum/oil ratio (Rc, normalized gum to oil concentration) to obtain the ratio at which the oil coverage is optimal. As shown

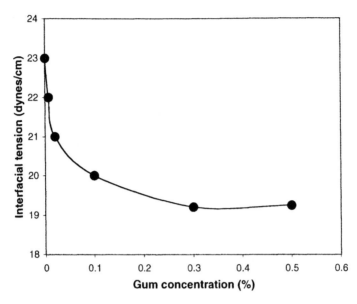

**Figure 5.14** Interfacial tension of *n*-tetradecane-water in the presence of tripurified fenugreek gum. Reprinted from Reference [28] with permission. © 1997 Academic Press Ltd., London, UK.

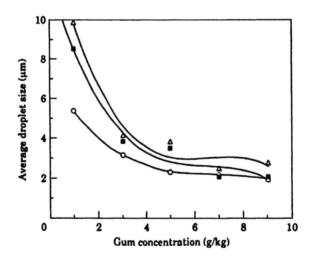

**Figure 5.15** Average droplet size of 50 g/kg *n*-tetradecane-in-water emulsion stabilized with fenugreek gum as emulsifier, at different aging times. (O) = 0 hr; (■) = 1 hr (△) = 168 hr. Reprinted from Reference [28] with permission. © 1997 Academic Press Ltd., London, UK.

244

TABLE 5.3. Effect of Gum and Oil Concentration ($R_c$ Index) on the Viscosity and Stability to Flocculation of O/W Emulsions.

| Gum Conc. %(w/w) | Oil Conc. (%) | (Gum/Oil) × 100 $R_c$ | Viscosity (Pas) | Flocculation |
|---|---|---|---|---|
| 0.1 | 5.00 | 2.00 | 27 | High |
| 0.2 | 1.66 | 12.05 | >25 | None |
| 0.3 | 1.66 | 18.07 | >25 | None |
| 0.3 | 2.14 | 14.02 | 30 | None |
| 0.3 | 5.00 | 6.00 | 48 | Medium |
| 0.3 | 6.42 | 4.67 | 55 | Medium |
| 0.5 | 2.77 | 18.05 | 67 | None |
| 0.5 | 4.16 | 12.02 | 100 | None |
| 0.5 | 5.00 | 10.00 | 115 | Small |
| 0.7 | 7.14 | 7.00 | 185 | Medium |
| 0.7 | 3.80 | 18.42 | 285 | None |
| 0.7 | 5.00 | 14.00 | 316 | None |
| 0.7 | 10.00 | 7.00 | 508 | Medium |
| 0.7 | 15.00 | 4.67 | 428 | Medium |
| 0.7 | 30.00 | 2.33 | 1440 | High |
| 0.9 | 5.00 | 18.00 | 474 | High |

Reprinted from Reference [28] with permission. © 1997 Academic Press Ltd., London, UK.

in Table 5.3, when Rc > 12, the emulsions are stable and free of flocculation for a long time. The critical value of Rc of fenugreek gum is smaller compared to that of LBG and guar (20 and 18, respectively), indicating that fenugreek gum has a more efficient interfacial coverage, therefore, it is concluded that fenugreek gum is more efficient than other galactomannans in stabilizing oil-in water emulsions [28]. From Table 5.3, it can also be observed that the viscosity does not play a significant role in stabilizing these emulsions. Emulsions with high viscosity but low Rc values (smaller than 12) and partial coverage showed no long-term stability and high flocculation. In contrast, emulsions with Rc > 12 values (full coverage), even with low viscosities, exhibited high stability to coalescence and to flocculation [28]. When the percentage of gum in the emulsifier phase is plotted against gum concentration (Figure 5.16), a remarkable efficiency is observed, and it reaches a maximum of 60% of the added gum adsorbed at 10% oil and 0.7% gum. The percentage of gum adsorbed was also plotted against the oil concentration, as shown in Figure 5.17. A maximum adsorption is obtained at 15% (w/w) oil, and as the oil concentration increased, the adsorption decreased, which is similar to some surface active proteins [28]. In summary, fenugreek gum is a better emulsifier/stabilizer for oil/water emulsion compared to other galactomannans. Most of the droplets of the oil/water emulsions were in the range of 2–5 μm (over 75% by volume or number) at 0.3% (w/w) of fenugreek gum level. Fenugreek gum stabilized emulsions could be aged for over four weeks at temperature ranges of 4–50°C without any significant size change,

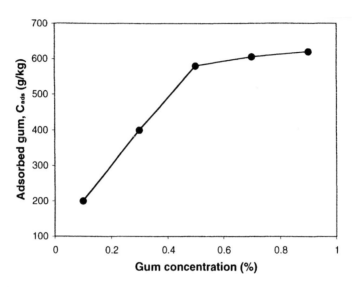

**Figure 5.16** Plot of the amount of fenugreek adsorbed on the *n*-tetradecane droplets as a function of gum concentration in the emulsification process. (Oil phase was 10%, w/w.) Reprinted from Reference [28] with permission. © 1997 Academic Press Ltd., London, UK.

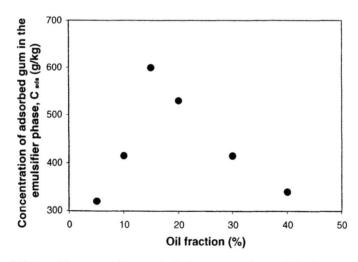

**Figure 5.17** Plot of the amount of fenugreek adsorbed on *n*-tetradecane oil droplets as a function of oil concentration in the emulsification system. (Gum concentration was 0.7%, w/w.) Reprinted from Reference [28] with permission. © 1997 Academic Press Ltd., London, UK.

which indicates that only slight coalescence took place [23]. Recent studies carried out in our laboratory confirmed that fenugreek gum is one of the best stabilizers in oil/water emulsions [24]. The excellent emulsifying properties exhibited by fenugreek gum will make fenugreek gum most useful in salad dressings, ice creams and other food emulsion systems.

## 4. WATER-SOLUBLE SOYBEAN POLYSACCHARIDES

Soybean, a yellow seed that originated from China, is a base of many food products in Oriental countries because of its high nutritional value [34–38]. One of the by-products of soybean processing is the residue produced during the making of tofu and soy protein, which are mainly composed of cell wall materials of soybean cotyledons. It has been reported that cotyledon cell walls contain substantial amounts of water-soluble polysaccharides [39,40]. Several acidic polysaccharide fractions and galactomannans have also been extracted from soybean hulls, and some of the fractions are extractable with cold water and/or warm water [41]. As a dietary fiber, soybean polysaccharides exhibited strong beneficial effects on gastrointestinal functions, nutrient balance, glucose tolerance and serum lipids in humans and in animal trials [42]. The water-soluble fraction of soybean polysaccharides could also be used as a stabilizer and/or thickening agent in the food industry due to the availability of the raw material. However, much research is needed for understanding the functional properties of these materials.

### 4.1. EXTRACTION AND STRUCTURES OF WATER-SOLUBLE SOYBEAN POLYSACCHARIDES

Sequential extraction of soybean hulls afforded a number of polysaccharide fractions. The extraction procedure and structures of these fractions were reported by Aspinall and coworkers in the 1960s [39–41,43–45]. The extraction of soybean hulls with water at room temperature and 60°C furnishes two galactomannans, galactomannan I and II, respectively, which differed only in the proportions of the two constituent sugars [45]. Galactomannan I and II were purified after precipitation of the contaminating acidic polysaccharides with copper salt. The hydrolysis of the two polysaccharides gave galactose and mannose in the ratio of 1:1.4 and 1:2.35, respectively. Methylation analysis and partial hydrolysis established that the two galactomannans contain linear chains of $(1\rightarrow4)$-linked β-D-mannopyranosyl residues to which different proportions of α-D-galactopyranosyl residues are attached as single unit side chains by $(1\rightarrow6)$-linkages. The water-extracted soybean hulls were further extracted with hot aqueous ammonium oxalate to give acidic polysaccharide III, which was comprised mainly of galacturonic acid with galactose, arabinose, xylose, fucose and rhamnose [41]. The hull residues were then delignified and further extracted with an aqueous solution of ethylenediaminetetra-acetic acid disodium salt to give another acidic polysaccharide IV, which gave a similar mixture of sugars to that of acidic polysaccharide III. Additional extraction of the hull residues with 10% potassium hydroxide gave a hemicellulose A fraction, which precipitated on acidification of the extract, and a hemicellulose B fraction, which precipitated upon addition of ethanol [41]. Hemicellulose A contained a xylose as the main constitute sugar, and a pure xylan was isolated after precipitation from alkaline solution as the insoluble copper complex. Partial acid hydrolysis of the xylan gave a mixture of acidic sugars among which 2-$O$-(glucopyranosyluronic acid)xylose was the major component [41]. Methylation analysis showed that the polysaccharide contained a backbone chain of (1→4)-linked β-D-xylopyranose residues to which a small proportion of D-glucuronic acid residues were attached as side chains by (1→2) linkage [41]. The structure of hemicellulose B is similar to that of hemicellulose A.

Aspinall and coworkers also extracted a neutral and acidic polysaccharide from soybean cotyledon meals [39,40,43]. Defatted soybean cotyledon meal was first extracted with boiling ethanol-water (4:1) to remove soluble sugars and to inactivate enzymes and then extracted with cold 0.2% sodium hydroxide to remove most of the proteins [39]. Polysaccharides were extracted from the residue meal with 2% ethylenediaminetetra-acetic acid disodium salt at 85–90°C. The extracted polysaccharides were fractionated on a DEAE-cellulose column. A neutral fraction was obtained in the elution of the starting buffer (0.025 M phosphate, pH 6), while an acidic polysaccharide was eluted by 0.25 and 0.5 M of phosphate buffer. The hydrolysis of the neutral polysaccharide gave arabinose and galactose in a ratio of 1:2.5. Mild acid hydrolysis of the arabinogalactan resulted in the preferential liberation of arabinose, suggesting that the residues of this sugar are probably present in the furanosyl form and at the exterior chains of the molecular structure. The methylation results indicated that the arabinose residues arise from an arabinogalactan rather than from an accompanying arabinan because of the absence of evidence for branching through arabinose residues [39]. The structures of the acidic preparations were partially revealed as complex pectic polysaccharides by characterization of oligosaccharides obtained from partial acidic hydrolysis and from enzyme hydrolysis [39].

Recently, Furuta and Maeda reported a procedure to extract the water-soluble polysaccharides from commercial soybean residue [46]. The commercial soybean residue (500 g) is mixed with distilled water (20 times its weight), and the pH of the mixture is adjusted to pH 5 with HCl and heated at 120°C for 1.5 hr in an autoclave. After cooling to room temperature, the suspension is centrifuged at 10,000 × g for 30 min, and the residue is washed with distilled water (twice its weight). The supernatants are combined and freeze-dried to produce 45.3%

TABLE 5.4. **Chemical and Sugar Compositions of Water-Soluble Soybean Polysaccharides (SSPS).**

| Sugar Composition (Weight Ratio) | Content (%) |
|---|---|
| Uronic acid | 23.4 |
| Galactose | 41.5 |
| Arabinose | 21.4 |
| Xylose | 5.6 |
| Fucose | 3.5 |
| Rhamnose | 2.5 |
| Glucose | 2.1 |
| Total Sugar | 83.6 |
| Ash | 5.3 |
| Crude Protein | 4.7 |
| Yield | 45.3 |

Reprinted from Reference [46] with permission from Elsevier Science.

of water-soluble soybean polysaccharides (SSPS) based on the starting residue [46]. The SSPS is composed of 83.6% total sugar, 5.3% ash and 4.7% crude protein. The sugar composition of SSPS is 41.5% galactose, 23.4% uronic acid, 21.4% arabinose with some minor amounts of xylose (5.6%), fucose (3.5%), rhamnose (2.5%) and glucose (2.1%), as shown in Table 5.4. SSPS mainly consists of four molecular fractions that correspond to four molecular weights (542,000, 150,000, 24,900 and 4700, respectively) based on pullulan standards [46], as shown in Figure 5.18.

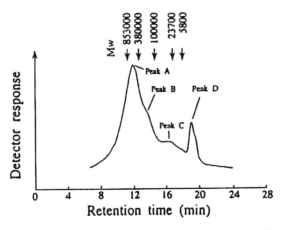

**Figure 5.18** Molecular weight distribution of water-soluble soy polysaccharides (SSPS) determined by GPC. Reprinted from Reference [46] with permission from Elsevier Science.

## 4.2. PHYSICAL PROPERTIES OF SOYBEAN POLYSACCHARIDES

The viscosity of SSPS increased with concentration, which is similar to pullulan and gum arabic, as shown in Figure 5.19. Although most of the polymers are in the high molecular weight region (Figure 5.18), its rheological response is rather weak: at 10% (w/w) polymer concentration, it exhibits Newtonian flow behavior in which the viscosity is independent of the shear rate. However, when SSPS is examined at 20% (w/w) solution, it exhibited shear thinning flow behavior (Figure 5.20). Furuta and Maeda also examined the effects of salts, sugar and pH on the viscosity [46]. It was found that the addition of three salts, NaCl, CaCl$_2$ and KCl did not have any significant effect on the viscosity over a wide range of salt concentrations examined; however, the viscosity of 10% SSPS increased steadily with the increase of sugar concentration, as demonstrated in Figure 5.21. The viscosity of SSPS solution tends to decrease with the decrease in pH, and the process is reversible (Figure 5.22). The viscosity of 10% SSPS solution is highly temperature dependent (Figure 5.23). The decrease of viscosity with increase of temperature is also pH dependent (Figure 5.24). Lower pH and high temperature resulted in the degradation of the polymer as indicated by a significant decrease in viscosity [46]. 10% SSPS solution prepared at pH 3 with 60% sucrose did not undergo gelation, indicating that it is different from high methoxyl pectins [46]. In summary, SSPS exhibited liquid fluid properties at conditions similar to those used for food and beverage sterilization and storage, therefore, it could be used in the food industry as a thickening and stabilizing agent. Sievert and coworkers compared the addition of soy polysaccharides and wheat bran in a breadmaking study [47]. 10% of soy polysaccharides had less effect than wheat bran on color and surface smoothness of Chinese steamed bread; meanwhile, it exerted a detrimental effect on volume

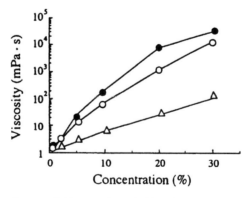

**Figure 5.19** Effect of water-soluble soy polysaccharides (SSPS) concentration on viscosity at a shear rate of 129/s and at 20°C. (O) SSPS, (●) pulullan, (△) gum arabic. Reprinted from Reference [46] with permission from Elsevier Science.

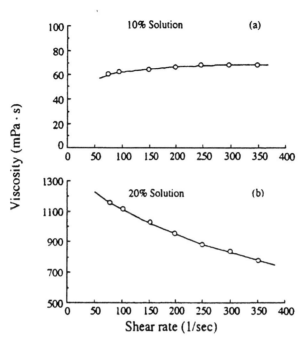

**Figure 5.20** Effect of shear rate on viscosity of 10 and 20% water-soluble soy polysaccharides (SSPS) solutions at 20°C. Reprinted from Reference [46] with permission from Elsevier Science.

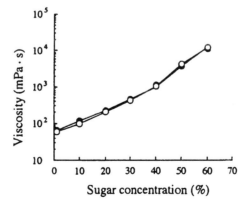

**Figure 5.21** Effect of sugar on viscosity of a 10% SSPS solution at different concentrations of sugar at a shear rate of 129/s and at 20°C. O Non-adjusted pH, ● pH 3. Reprinted from Reference [46] with permission from Elsevier Science.

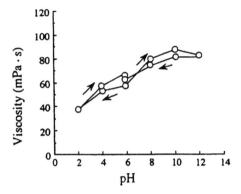

**Figure 5.22** Effect of pH on viscosity of 10% water-soluble soy polysaccharides (SSPS) solution. 50% NaOH or 12 N hydrochloric acid was added to increase or decrease the pH of the 10% SSPS solution to pH 2.0 or 12.0, respectively, and the viscosity was measured at a shear rate of 129/s and at 20°C. Reprinted from Reference [46] with permission from Elsevier Science.

and texture. Soy polysaccharides also exhibited superior properties regarding volume, color and crumb grain when added to a Japanese sponge cake at 4% [47]. Further extensive research is needed to fully understand the structural and functional properties of soybean polysaccharides that might be a viable food hydrocolloid for the food industry.

## 5. CORN FIBER GUM

Corn fiber, composed mostly of kernel hull or the pericarp, is a by-product of the corn wet milling industry, and over 4 million tons per year are produced in

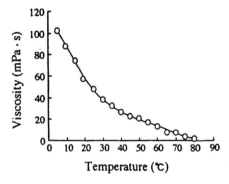

**Figure 5.23** Effect of temperature on viscosity of a 10% water-soluble soy polysaccharides (SSPS) solution at a shear rate of 129/s and heating rate of 1°C/min. Reprinted from Reference [46] with permission from Elsevier Science.

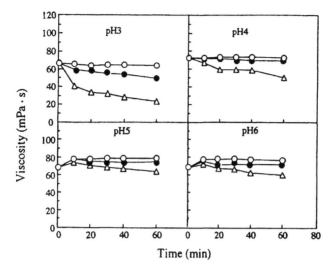

**Figure 5.24** Effect of heating duration time at adjusted temperature 70, 100, 120°C under acidic conditions on viscosity of a 10% SSPS solution. Viscosity was measured at a shear rate of 129/s after cooling to 20°C. O 70°C, ● 100°C, △ 120°C. Reprinted from Reference [46] with permission from Elsevier Science.

the U.S. [48]. Corn fiber consists of 40–48% hemicellulose, 20–22% cellulose, 9–23% starch, 10–13% proteins, 3–4% uronic acid, 2–3% oil and ~2% ash [49]. The hemicellulose portion of corn fiber can be extracted with alkaline solutions to produce water-soluble gum-corn fiber gum (CFG). The chemical structure, physical properties and application of corn fiber gum have been reviewed by Whistler [50]; however, all of the literature cited in this review was published in the 1950s and 1960s, which may reflect the lack of interest in the area for over two to three decades. A recent revisit to corn fiber gum provided new information on the extracting process and characterization of physicochemical properties of this material [49,51]. This section intends to review corn fiber gum consulting up-to-date literature.

## 5.1. EXTRACTION, MONOSACCHARIDE COMPOSITION AND STRUCTURES

Earlier reports discussed extracting corn fiber gum by first boiling the corn fiber in alkaline solutions, such as sodium hydroxide or sodium carbonate at pH 10.5–11.5 for 1 hr and then filtering or centrifuging [52]. The supernatant was acidified to pH 4.0 with HCl, and the gum was precipitated with isopropanol [52]. Corn fiber extracted with lime water gave a gum product with lighter color that was more acceptable for further decolorization [49,50]. It appears that the extracting solvents have a major effect on the yield and composition of corn

TABLE 5.5. **Chemical and Sugar Compositions of Corn Fiber Gum Isolated by Calcium Hydroxide and Ammonium Hydroxide Extractions.**

| Preparation | 2C-A | 15NH-A | 15NH-BK | 15NH-BKC |
|---|---|---|---|---|
| 1st extract | 2% Ca(OH)$_2$ | 15% (NH$_4$)OH | 15% (NH$_4$)OH | 15% (NH$_4$)OH |
| 2nd extract | | | 0.5% KOH | 0.5% KOH |
| 3rd extract | | | | 2.5% Ca(OH)$_2$ |
| Total neutral sugars (%)[a] | 80.1 | 64.5 | 68.8 | 90 |
| Distribution of neutral sugars (%) | | | | |
| arabinose | 36.3 | 31.6 | 33.8 | 35.7 |
| xylose | 56.6 | 58.1 | 57.8 | 56.8 |
| galactose | 7.3 | 8.6 | 7.4 | 6.6 |
| glucose | 0.5 | 1.6 | 0.6 | 0.6 |
| Uronic acids (%)[a] | 7.7 | 5.5 | 8.0 | 6.9 |
| Protein (%)[a] | 2.1 | 7.2 | 4.6 | 2.4 |
| Recovery (%)[b] | 14.0 | 11.1 | 8.2 | 5.1 |

[a]Expressed percent of dry weight of xylan recovered.
[b]As percent weight of initial corn fiber based on orcinol sugar content of corn fiber and recovered xylan.
Reprinted from Reference [49] with permission from the *Journal of Agricultural and Food Chemistry.* © 1998 American Chemical Society.

fiber gum, as shown in Table 5.5 [49]. A recent report on the extraction of corn fiber gum includes the following steps, as demonstrated in Figure 5.25 [48,51]. The corn fiber is first destarched by incubating with α-amylase. The destarched corn fiber is extracted with alkaline solution and water at 100°C for 1 hr. The alkali extract is treated with H$_2$O$_2$ to remove the color. After adjusting the pH to 4.0–4.5, hemicellulose A (precipitate) is removed by vacuum filtration through a Celite filter aid (or by centrifugation at 10,000×g for 10 min). The CFG (hemicellulose B) is precipitated with two volumes of 95% ethanol. The precipitate is stirred in 95% ethanol for 5 min, isolated by filtration, air-dried in a fume hood, then dried in a vacuum oven at 50°C for 1 hr. The CFG is then converted to a fine white powder with a grinder [48]. The removal of starch and water-insoluble hemicellulose A was considered essential for obtaining a clear solution of CFG [51]. The alkaline hydrogen peroxide bleaching was conducted at ambient temperature for 2 hr to obtain optimal effect. The yield and compositions of corn fiber gum extracted under different conditions and treated with H$_2$O$_2$ are shown in Table 5.6. It appeared that the yield of CFG was correlated with the pH of the extraction solvent. The pH of Ca(OH)$_2$ solution was 9.8 which yielded 21% of CFG; the pH of NaOH was 11.1, and it gave a yield of 40%. When the mixture of NaOH and Ca(OH)$_2$ was used at a 1:1 ratio, the pH was 10.3 which afforded a yield of 27% (Table 5.6). The neutral sugar

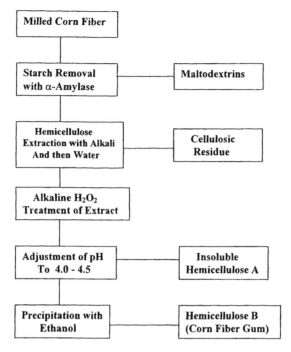

**Figure 5.25** Flow diagram for production of corn fiber gum (hemicellulose B) from milled corn fiber. Reprinted from Reference [48].

and uronic acid levels of CFG did not change significantly under different extraction conditions, however, the molecular weight of corn fiber gums changed significantly under different extraction conditions. For example, The MW of NaOH-extracted CFG was $3.94 \times 10^5$, while the MW of Ca(OH)$_2$-extracted material was only $2.78 \times 10^5$. It appeared that extreme alkaline conditions resulted in the liberation of higher molecular species of hemicellulose B; this observation is in agreement with an earlier report by Saulnier and coworkers [53]. In this study, a hemicellulose extracted from corn bran with 0.5 M NaOH at 30°C for 2 hr had a MW of $2.7 \times 10^5$; when the residue material was extracted with 1.5 M KOH for 2 hr at 100°C, the obtained hemicellulose had a MW of $3.7 \times 10^5$. The Ca(OH)$_2$-extracted CFG appeared very white compared to that of NaOH-extracted CFG, which was off-white [48].

The range of monosaccharide composition of corn fiber from different groups was D-xylose (48–54%), L-arabinose (33–35%), galactose (5–11%) and D-glucuronic acid (3–6%) [53–55]. The structure of CFG is featured by a $(1\rightarrow4)$-linked-β-D-xylopyranose backbone heavily substituted with side groups. The side chains consist mostly of α-L-arabinofuranosyl residues mainly at *O*-3 [56], but also at *O*-2 of the xylosyl residues with xylose and galactose side

TABLE 5.6. Chemical and Sugar Compositions (%) and Some Physical Properties of Corn Fiber Gum Extracted Under Different Conditions and Treated with $H_2O_2$.

|  | NaOH | Ca(OH)$_2$ | 1:1 |
|---|---|---|---|
| Yield[a] | 40 | 21 | 27 |
| Ash | 2.15 | 2.65 | 2.42 |
| N | 0.151 | 0.135 | 0.183 |
| Ca | 0.30 | 0.56 | 0.63 |
| Na | 0.52 | 0.042 | 0.13 |
| Mr($\times$105) | 3.94 | 2.78 | 3.03 |
| Whiteness index[b] | 38.9 | 50.3 | 49.4 |
| Sugar composition[c] |  |  |  |
| Arabinose | 39.4 | 37.7 | 40.8 |
| Xylose | 48.1 | 49.8 | 49.5 |
| Galactose | 8.4 | 7.5 | 5.4 |
| Glucose | 0.8 | 1.0 | 0.8 |
| Glucuronic acid | 4.2 | 4.9 | 4.3 |

[a]Dry weight basis. Starting corn fiber had 5.8% moisture.
[b]Standard = 83.2.
[c]Relative percentages.
Reprinted from Reference [48].

chains [53]. The galactose and xylose residues are linked to the arabinofuranosyl branches [56]. The D-glucuronic acid residues are linked at the O-2 of the backbone xylosyl residues [57].

## 5.2. PHYSICAL PROPERTIES AND APPLICATIONS OF CORN FIBER GUM

Corn fiber gum dissolves in cold water, hot water, acidic and alkaline pHs. It is also soluble in glycerol and in 55% aqueous ethanol, suggesting that the CFGs contain relatively small molecular weight polysaccharides. Corn fiber gum resembles gum arabic in its solubility characteristics; a water solution of up to 30–40% of solids is readily prepared [50]. However, the viscosity of corn fiber gum is most nearly like gum karaya, which shows a sharp increase in viscosity with increasing polymer concentration (Figure 5.26) [50]. Figure 5.27 demonstrates the effects of temperature and salts on viscosity of corn fiber gum solutions. A linear decrease of viscosity with an increase of temperature is observed on the semi-log plot. In addition, the viscosities of 5% and 10% solutions of CFG are not significantly affected by the presence of NaCl and $CaCl_2$ at the 0.1 M level [48]; which is in agreement with Whistler that 10% CFG solution was not significantly affected by a number of ordinary salts at 0.125% concentration [50]. However, the viscosity of 10% CFG solution was increased from 19.5 poises to 315 poises in 0.125% of sodium tetraborate solution [50]. Corn

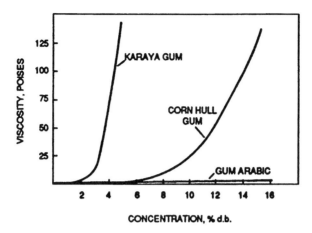

**Figure 5.26** Viscosity versus concentration of corn fiber gum, gum arabic and gum karaya. Brookfield viscometer, No. 4 spindle, 60 rpm. Reprinted from Reference [50] by permission of Academic Press, Inc., © 1993.

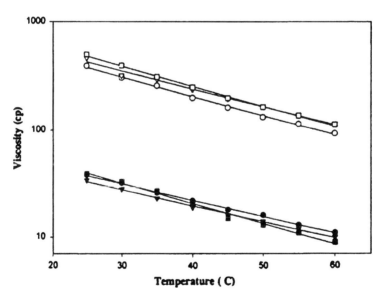

**Figure 5.27** Viscosities (cP) vs. temperature for 5 and 10% corn fiber gum solutions at pH 5.5 in water (● 5%, O 10%), 100 mM NaCl (▼ 5%, □ 10%) and 100 mM $CaCl_2$ (■ 5%, ❑ 10%). Reprinted from Reference [48].

fiber gum also exhibited good film-making properties [50]. Further research is needed to fully understand the physicochemical and functional properties of corn fiber gum and to turn the massive amounts of by-products from the corn industry into a valued polysaccharide gum.

## 6. REFERENCES

1. Glicksman M. *Gum Technology in the Food Industry*. New York: Academic Press; 1969.

2. Leeds A. R. *Dietary Fiber Perspectives*. London: John Libbey; 1985.

3. Matheson H. B., Colon I. S., Story J. A. Cholesterol 7-alpha-hydroxylase activity is increased by dietary modification with psyllium hydrocolloid, pectin, cholesterol and cholestyramine in rats. *J. Nutrition* 1995; 125:454–8.

4. Turley S. D., Dietschy J. M. Mechanisms of LDL-cholesterol lowering action of psyllium hydrophillic mucilloid in the hamster. *Biochim. Biophys. Acta* 1995; 1255:177–84.

5. Hefti M. C. Separation of nonsugar polysaccharide from plant seeds of fruits. *Chem. Abstr.* 1954; 51:8459. Ital. Patent 498,739.

6. Near H. B., Pacini A. J., Crosley R. W., Gerth M. M., Breidigam F. T., Kelly J. D. Extracting mucinous substances from plant materials such as seeds of *Plantago psyllium*. *Chem. Abstr.* 1935; 29:6674. U.S. Patent 2,010,880.

7. McCredie R. J., Whistler R. L. In *Industrial Gums*, first edition. Whistler R. L., ed. New York: Academic Press; 1959. p. 433.

8. Montague J. E. *Psyllium Seed: The Latest Laxative*. New York: Montague Hospital for Intestinal Ailments; 1932.

9. Tomoda M., Ishikawa K., Yokoi M. Plant mucilages. 29. Isolation and characterization of a mucous polysaccharide, plantago-mucilage-A, from the seeds of *Plantago-major* var asiatica. *Chem. Pharm. Bull.* 1981; 29:2877–84.

10. Kennedy J. F., Sandhu J. S., Southgate D. A. Structural data for the carbohydrate of ispaghula husk ex *Plantago ovata* Forsk. *Carbohydr. Res.* 1979; 75:265–74.

11. Haque A., Richardson R. K., Morris E. R., Dea I. C. M. Xanthan-like "weak gel" rheology from dispersions of ispaghula seed husk. *Carbohydr. Polym.* 1993; 22:223–32.

12. Tomoda M., Gonda R., Sakabe H., Shimada K., Shimizu N. Plant mucilages. 34. The location of *O*-acetyl groups and the structural features of plantago-mucilage-A, the mucous polysaccharide from the seeds of *Plantago-major* var asiatica. *Chem. Pharm. Bull.* 1984; 32: 2182–6.

13. Haque A., Morris E. R., Richardson R. K. Polysaccharide substitutes for gluten in non-wheat bread. *Carbohydr. Polym.* 1994; 25:337–44.

14. Clark A. H., Ross-Murphy S. Structural and mechanical properties of biopolymer gels. *Advan. Polym. Sci.* 1987; 83:57–192.

15. Haque A. H., Morris E. R., Richardson R. K. *Frontiers in Carbohydrate Research 3*. Millane T. E., BeMiller J. N., Chandrasekaran R., eds. Orlando, FL: Academic Press, Inc.; 1993.

16. Anderson J. W., Zettwoch N., Feldman T., Tietyenclark J., Oeltgen P., Bishop C. W. Cholesterol-lowering effects of psyllium hydrophilic mucilloid for hypercholesterolemic men. *Arch. Intern. Med.* 1988; 148:292–6.

17. Turley S. D., Daggy B. P., Dietschy J. M. Cholesterol-lowering action of psyllium mucilloid in the hamster—sites and possible mechanisms of action. *Metabolism* 1991; 40:1063–73.

18. Bone J. N., Rising L. W. An *in vitro* study of various commercially available bulk-type laxatives. *J. Amer. Pharm. Ass., Sci. Ed.* 1954; 43:102–6.

19. MacDonald N. S., Nusbaum R. E., Ezmirlian F., Barbera R. C., Alexander G. V., Spain P., Round D. E. *J. Pharmacol. Exp. Ther.* 1952; 104:348.

20. Taylor W. G., Zaman M. S., Mir A., Mir P. S., Acharya S. N., Mears G. J., Elder J. L. Analysis of steroidal sapogenins from amber fenugreek (*Trigonella foenum-graecum*) by capillary gas chromatography and combined gas chromatography/mass spectrometry. *J. Agric. Food Chem.* 1997; 45:753–9.

21. Sharma R. D. Effect of fenugreek seeds and leaves on blood glucose and serum insulin responses in human subjects. *Nutrition Res.* 1986; 6:1353–64.

22. Madar A., Shomer I. Polysaccharide composition of a gel fraction derived from fenugreek and its effect on starch digestion and bile acid absorption in rats. *J. Agric. Food Chem.* 1990; 38:1535–9.

23. Garti N. Hydrocolloids as emulsifying agents for oil-in-water emulsions. *J. Dispersion Sci. and Techn.* 1999; 20:327–55.

24. Huang X., Cui W., Kakuda Y. Systematic evaluation of hydrocolloid gums as stabilizers in oil/water emulsion. *Food Hydrocolloid*, 2000; in press.

25. Rao P. U., Sesikeran B., Rao P. S., Naidu A. N., Rao V. V., Ramacjandran E. P. Short term nutritional and safety evaluation of fenugreek. *Nutrition Res.* 1996; 16:1495–505.

26. Reid J. S., Meier H. Formation of reserve galactomannan in the seeds of *Trigonella foenumgraecum*. *Phytochemistry* 1970; 9:513–20.

27. Andrews P., Hough L., Jones K. N. Mannose containing polysaccharides. Part II. The galactomannan of fenugreek seed. *J. Amer. Chem. Soc.* 1952; 2744–50.

28. Garti N., Madar Z., Aserin A., Sternheim B. Fenugreek galactomannans as food emulsifiers. *Food Sci. and Techn.* 1997; 30:305–11.

29. Maier H., Anderson M., Kari C., Magnuson K., Whistler R. L. Guar, locust bean, tata, and fenugreek gums. In *Industrial Gums*, third edition. Whistler R., La B. J. N., eds. San Diego, London: Academic Press, Inc.; 1993. pp. 181–226.

30. Garti N., Reichman D. Surface properties and emulsification activity of galactomannans. *Food Hydrocolloids* 1994; 8:155–73.

31. Gaonkar A. G. Surface and interfacial activities and emulsion characteristics of some food hydrocolloids. *Food Hydrocolloids* 1991; 5:328–37.

32. Dickinson E., Galazka V. B., Anderson D. M. W. Emulsifying behaviour of acacia gums. In *Gums and Stabilisers for the Food Industry 5*. Phillips G. O., Wedlock D. J., Willians P. A., eds. Oxford: IRL Press; 1990. pp. 41–4.

33. Benjamins J., De Feijter J. A., Evans M. T. A., Grahan D. E., Phillips M. C. Dynamic and static properties of proteins adsorbed at the air/water interface. *Faraday Discussions of the Chemical Society* 1975; 59:218–29.

34. Steinki F. H. Nutritional value of soybean protein foods. In *New Protein Foods in Human Health: Nutrition, Prevention, and Therapy*. Waggle D. H., Steinke F. H., Volgarev M. N., eds. Boca Raton, FL: CRC Press; 1992. pp. 59–66.

35. Fernando S. M., Murphy P. A. HPLC determination of thiamin and riboflavin in soybeans and tofu. *J. Agric. Food Chem.* 1990; 38:163–7.

36. Wang G., Kuan S. S., Francis O. J., Ware G. M., Carman A. S. A simplified HPLC method for the determination of phytoestrogens in soybean and its processed products. *J. Agric. Food Chem.* 1990; 38:185–90.

37. Barnes S., Kirk M., Coward L. Isoflavones and their conjugates in soy foods: extraction conditions and analysis by HPLC-mass spectrometry. *J. Agric. Food Chem.* 1994; 42:2466–74.

38. Henley E. C., Steinke F. H., Waggle D. H. Nutritional value of soy protein products. *Proc. World Conf. Oilseed Technol. Util. 1992*. Champaign: IL: American Oil Chemists Society; 1993. pp. 248–56.

39. Aspinall G. O., Begbie R., Hamilton A., Whyte J. N. C. Polysaccharides of soy-beans. Part III. Extraction and fractionation of polysaccharides from cotyledon meal. *J. Chem. Soc.* 1967; C:1065–70.

40. Aspinall G. O., Cottrell I. W., Egan S. V., Morrison I. M., Whyte J. N. Polysaccharides of soy-beans. Part IV. Partial hydrolysis of the acidic polysaccharide complex from cotyledon meal. *J. Chem. Soc.* 1967; C:1072–80.

41. Aspinall G. O., Hunt K., Morrison I. M. Polysaccharides of soy-beans. Part II. Fractionation of hull cell-wall polysaccharides and the structure of a xylan. *J. Chem. Soc* 1966; C: 1945–9.

42. Tsai A. C., Mott E. L., Owen G. M., Bennick M. R., Lo G. S., Steinke F. H. Effects of soy polysaccharide on gastrointestinal functions, nutrient balance, steroid excretions, glucose tolerance, serum lipids, and other parameters in humans. *Am. J. Clin. Nutr.* 1983; 38:504–11.

43. Aspinall G. O., Cottrell I. W. Polysaccharides of soybeans. VI. Neutral polysaccharides from cotyledon meal. *Can. J. Chem.* 1971; 49:1019–22.

44. Aspinall G. O., Hunt K., Morrison I. M. Polysaccharides of soy-beans. Part V. Acidic polysaccharides from the hulls. *J. Chem. Soc.* 1967; C:1080–6.

45. Aspinall G. O., Whyte J. N. C. Polysaccharides of soy-beans. Part I. Galactomannans from the hulls. *J. Chem. Soc.* 1964; 972:5058–63.

46. Furuta H., Maeda H. Rheological properties of water-soluble soybean polysaccharides extracted under weak acidic condition. *Food Hydrocolloids* 1999; 13:267–74.

47. Sievert D., Pomeranz Y., Abdelrahman A. Functional-properties of soy polysaccharides and wheat bran in soft wheat products. *Cereal Chem.* 1990; 67:10–3.

48. Doner L. W., Chau H. K., Fishman M. L., Hicks K. B. An improved process for isolation of corn fiber gum. *Cereal Chem.* 1998; 75:408–11.

49. Hespell R. B. Extraction and characterization of hemicellulose from the corn fiber produced by corn wet-milling processes. *J. Agric. Food Chem.* 1998; 46:2615–9.

50. Whistler R. L. Hemicelluloses. In *Industrial Gums*. Whistler R. L., BeMiller J. N., eds. Orlando, FL: Academic Press, Inc.; 1993. pp. 295–308.

51. Doner L. W., Hicks K. B. Isolation of hemicellulose from corn fiber by alkaline hydrogen-peroxide extraction. *Cereal Chem.* 1997; 74:176–81.

52. Wolf M. J., MacMasters M. M., Cannon. J. A., Rosewell E. C., Rist C. E. Preparation and some properties of hemicelluloses from corn hulls. *Cereal Chem.* 1953; 30:451–70.

53. Saulnier L., Peneau N., Thibault J. F. Variability in grain extract viscosity and water-soluble arabinoxylan content in wheat. *J. Cereal Sci.* 1995; 22:259–64.

54. Whistler R. L., BeMiller J. N. Hydrolysis of components from methylated corn fiber gum. *J. Am. Chem. Soc.* 1956; 78:1163–5.

55. Suguwara M., Suzuki T., Totsuka A., Takeuchi M., Ueki K. Composition of corn hull dietary fiber. *Starch/Staerke* 1994; 46:335–7.

56. Whistler R. L., Corbett W. M. Oligosaccharides from partial hydrolysis of corn fiber hemicellulose. *J. Am. Chem. Soc.* 1955; 77:6328–30.

57. Montgomery R., Smith F. Structure of corn hull hemicellulose. III. Identification of the methylated aldobiouronic acid obtained from methyl corn hull hemicellulose. *J. Am. Chem. Soc.* 1957; 79:695–7.

# Index